Comprehensive Gynecology and Obstetrics

This series presents the current and future perspectives of medical science in gynecology and obstetrics. The authors fully describe the current understanding of a disease including clinical features, imaging, pathology, and molecular biology, and also include the historical aspects and theories for exploring the etiology of the disease. Also, recent developments in diagnostic strategy, medical treatment, surgery, radiotherapy, prevention, and better health-care methods are clearly shown. Thus, each volume in the series focuses on the scientific basis for the pathogenesis of a disease and provides clinical applications that make it possible to offer personalized treatment for each patient. Over the past 20 years, physicians have been working to develop a standard treatment and publish clinical guidelines for a disease based on epidemiological evidence, mainly through the use of randomized clinical trials and meta-analyses. Recently, however, comprehensive genomic and genetic analyses have revealed the differences and variations in biological characteristics even among patients with the same diagnosis and have been focusing on personalized therapy. Now all physicians and patients are entering a new world of "precision medicine" through the use of genomic evidence. We are confident that readers will greatly benefit from the contents of the series with its purview of the exciting and promising future of gynecology and obstetrics.

Hidemichi Watari • Takashi Mitamura
Editors

Uterine Body Neoplasms

Diagnosis and Treatment

Springer

Editors
Hidemichi Watari
Department of Obstetrics and Gynecology
Hokkaido University Graduate School of Medicine
Sapporo, Hokkaido, Japan

Takashi Mitamura
Department of Obstetrics and Gynecology
Hokkaido University Graduate School of Medicine
Sapporo, Hokkaido, Japan

ISSN 2364-1932 ISSN 2364-219X (electronic)
Comprehensive Gynecology and Obstetrics
ISBN 978-981-95-0945-4 ISBN 978-981-95-0946-1 (eBook)
https://doi.org/10.1007/978-981-95-0946-1

This Springer imprint is published by the registered company Springer Nature Singapore Pte Ltd.
The registered company address is: 152 Beach Road, #21-01/04 Gateway East, Singapore 189721, Singapore

Preface

Endometrial cancer is the most prevalent in malignancies arising from female genital tract in Japan as well as in western countries. The endometrium, which is the origin of endometrial cancer, already harbors cancer-associated gene mutations due to several factors such as aging or menstruation, even though it looks histologically normal. Previous studies have identified driver genes in endometrial cancer and have shown that molecular subtype classifications reflect its biological characteristics more accurately than the conventional classification. Indeed, The Cancer Genome Atlas (TCGA) demonstrated that endometrial cancer was molecularly classified into four subtypes, and the patients' prognosis can be stratified. In 2020, the World Health Organization (WHO) revised its histological classifications of gynecological tumors and issued the WHO Classification of Tumours of Female Reproductive Organs 5th edition, in which TCGA molecular classification was introduced as WHO classification for endometrial cancer. In 2023, International Federation of Gynecology and Obstetrics (FIGO) proposed the revised staging system for endometrial cancer, in which molecular subtype classification is strongly encouraged and incorporated into staging system for the cases with early-stage disease. How to implement molecular classification and FIGO 2023 in clinical practice is the matter of discussion at this moment in the world.

Regarding treatments for endometrial cancer, minimally invasive surgery, including laparoscopic or robot-assisted surgery with or without sentinel node navigation, have become common for early-stage disease, and a variety of target drugs in single use or in combination with conventional cytotoxic agents have been developed for advanced or recurrent endometrial cancer. Based on molecular classification, de-escalation of treatment (omission of adjuvant therapy) would be recommended for *POLE* mutant endometrial cancer, but multidisciplinary treatments with the combination of complete surgical staging including systematic lymphadenectomy and more intensive drug treatments using immune checkpoint inhibitors should be recommended for advanced or recurrent cases with poor prognosis, which would be expected to provide better outcome of the patients with endometrial cancer. In the near future, we can expect that precision care will be indicated from the beginning of treatment based on the molecular classification and biomarker-driven patient selection through implementation of biomarker testing approved by the government as the companion diagnostics in clinical practice in Japan.

For this book, we asked gynecologic oncology experts in Japan to write original chapters. All authors are hoping and confident that readers will greatly benefit from the contents of this book. We sincerely hope that this book would become a must-have resource not only for gynecologic oncologists but also for many doctors, ranging from those who have just started engaging in research or clinical care to experienced gynecologists.

Sapporo, Hokkaido, Japan
April 2025

Hidemichi Watari
Takashi Mitamura

Contents

1 Overview on the Diagnosis and Treatment of Uterine Body Neoplasms: From Worldwide and Japanese Perspectives 1
Hidemichi Watari

2 Molecular Mechanism of Endometrial Carcinogenesis 11
Kotaro Takahashi and Kosuke Yoshihara

3 Hereditary Aspects of Endometrial Cancer 27
Takashi Mitamura

4 FIGO2023 Staging System and Molecular Classification for Endometrial Cancer 45
Hiroyuki Yamazaki and Hiroshi Asano

5 Artificial Intelligence in the Diagnosis of Endometrial Cancer 65
Yusuke Toyohara and Kenbun Sone

6 Lymphadenectomy for Endometrial Cancer 79
Yosuke Konno

7 Sentinel Node Navigation Surgery for Endometrial Cancer 93
Hideaki Yahata

8 Minimally Invasive Surgery for Endometrial Cancer 105
Yoshito Terai

9 Immune Checkpoint Inhibitors in Endometrial Cancer: Current Status and Future Perspectives 115
Junzo Hamanishi

10 Novel Treatment for Endometrial Cancer 133
Takashi Mitamura

11 Fertility-Preserving Treatment for Endometrial Cancer 151
Shoko Kitazawa, Kensuke Sakai, and Wataru Yamagami

12 Uterine Sarcoma 173
Hiroshi Asano

1 Overview on the Diagnosis and Treatment of Uterine Body Neoplasms: From Worldwide and Japanese Perspectives

Hidemichi Watari

1.1 Introduction

Endometrial cancer is the most prevalent among malignancies arising from the female genital tract. A total of 13,912 newly diagnosed cases have been registered according to the patient annual report for 2021 in Japan [1], and the number of patients has been increasing. Since around 70% of the patients with endometrial cancer are diagnosed as stage I disease, overall prognosis of endometrial cancer patients seems good. However, because the prognosis of advanced or recurrent cases is not always good, we, gynecologic oncologists, have been waiting for new treatment options to improve the outcome of the treatment for endometrial cancer patients.

I would expect that this book would highlight the difference of treatment strategy for endometrial cancer between western countries and Japan. First, postsurgical recurrent risk has been classified into three groups—low, intermediate, and high—after surgical staging in Japan (Fig. 1.1) according to the latest Japanese guideline (Japan Society of Gynecologic Oncology 2023 guidelines for treatment of uterine body neoplasm) [2]. In the European Society of Gynaecological Oncology (ESGO)/ European Society of Radiotherapy and Oncology (ESTRO)/European Society of Pathology (ESP)2021 guidelines for the management of patients with endometrial carcinoma, five prognostic risk groups have been defined: low, low-intermediate, high-intermediate, high, and advanced/metastatic (Table 1.1) [3]. The ESGO/ ESTRO/ESP guidelines consider molecular classification described below.

Second, regarding postsurgical adjuvant therapy, western countries routinely administer radiotherapy (brachytherapy or pelvic radiotherapy) depending on the risk of recurrence. In contrast, the Japanese guideline [2] has proposed/

H. Watari (✉)
Department of Obstetrics and Gynecology, Hokkaido University Graduate School of Medicine, Sapporo, Japan
e-mail: watarih@med.hokudai.ac.jp

H. Watari, T. Mitamura (eds.), *Uterine Body Neoplasms*, Comprehensive Gynecology and Obstetrics, https://doi.org/10.1007/978-981-95-0946-1_1

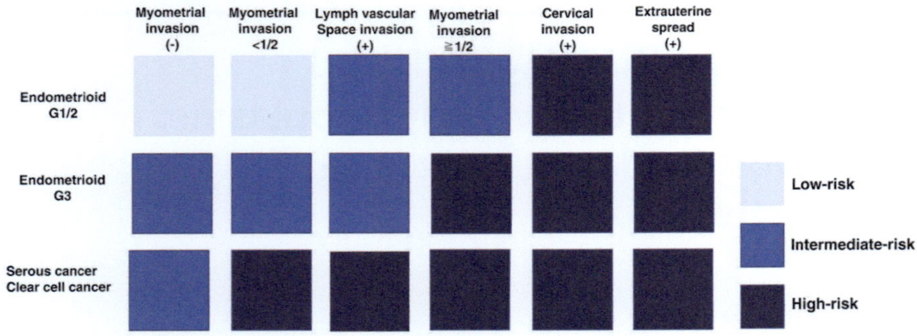

Fig. 1.1 Postsurgical recurrent risk classification in Japan

Table 1.1 Definition of prognostic risk groups in ESGO/ESTRO/ESP guidelines

Risk group	Molecular classification unknown	Molecular classification known
Low	· stage IA endometrioid+low-grade*+LVSI negative/focal	· stage I-II *POLE*mut endometrial carcinoma, no residual disease · stage IA **MMRd/NSMP** endometrioid carcinoma + low-grade*+LVSI negative/focal
Intermediate	· stage IB endometrioid+low-grade*+LVSI negative/focal · stage IA endometrioid+high-grade*+LVSI negative/focal · stage IA non-endometrioid (serous, clear cell, undifferentiated carcinoma, carcinosarcoma, mixed) without myometrial invasion	· stage IB **MMRd/NSMP** endometrioid carcinoma+low-grade*+LVSI negative/focal · stage IA **MMRd/NSMP** endometrioid carcinoma+high-grade*+LVSI negative/focal · stage IA **p53 abn** and/or non-endometrioid (serous, clear cell, undifferentiated carcinoma, carcinosarcoma, mixed) without myometrial invasion
High-intermediate	· stage I endometrioid+substantial LVSI, regardless of grade, depth of invasion · stage IB endometrioid+high-grade* regardless of LVSI status · stage II	· stage I **MMRd/NSMP** endometrioid carcinoma+substantial LVSI, regardless of grade, depth of invasion · stage IB **MMRd/NSMP** endometrioid carcinoma+high-grade* regardless of LVSI status · stage II **MMRd/NSMP** endometrioid carcinoma
High	· stage III-IVA with no residual disease · stage I-IVA non-endometrioid (serous, clear cell, undifferentiated carcinoma, carcinosarcoma, mixed) with myometrial invasion and with no residual disease	· stage III-IVA **MMRd/NSMP** endometrioid carcinoma with no residual disease · stage I-IVA **p53 abn** endometrial carcinoma with myometrial invasion, qith no residual disease · stage I-IVA **NSMP/MMRd** serous, undifferentiated carcinoma, carcinosarcoma with myometrial invasion, qith no residual disease
Advanced Metastatic	· stage III-IVA with residual disease · stage IVB	· stage III-IVA with residual disease of any molecular type · stage IVB of any molecular type

p53abn p53 abnormal, *MMRd* mismatch repair deficiency, *NSMP* nonspecific molecular profile, *POLE*mut polymerase ε mutated

[a]According to the binary FIGO grading, grade 1 and grade 2 carcinomas are considered as low-grade, and grade 3 carcinomas are considered as high-grade

recommended systemic chemotherapy, not radiotherapy, as adjuvant treatment for endometrial cancer at intermediate/high risk of recurrence (Fig. 1.2), because we, Japanese gynecologic surgeons, believe that extensive surgery including systematic pelvic lymphadenectomy can reduce the risk of local recurrence, not by adjuvant radiotherapy, and we aim to reduce the risk of distant failure by adjuvant chemotherapy.

Here, I would summarize the recommendations of treatment choice for endometrial cancer according to the latest Japanese guideline [2] and give an overview of the contents of this book, which would provide recent information and future perspectives on various aspects of the diagnosis and treatment of uterine body neoplasms by experts in the field of gynecologic oncology from Japan.

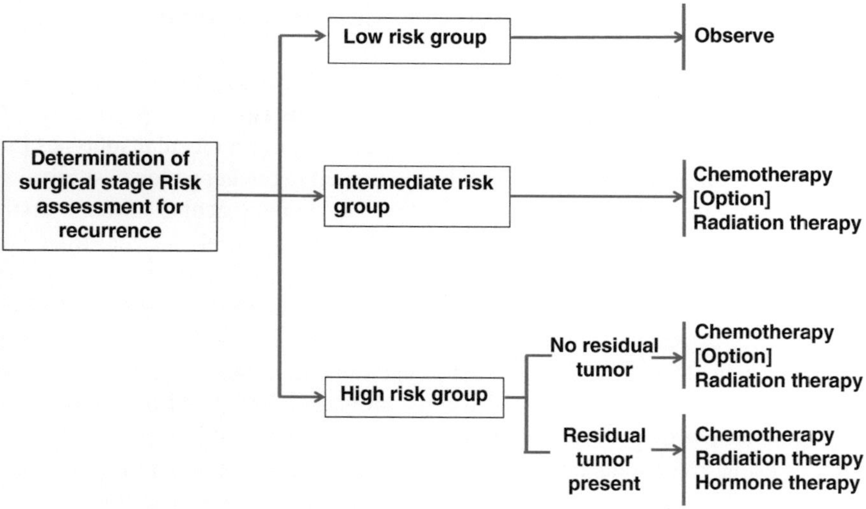

Fig. 1.2 Postoperative adjuvant treatment for endometrial cancer in Japan

1.2 Molecular Mechanism of Endometrial Carcinogenesis

Endometrial cancer arises from endometrial epithelium by the accumulation of pathogenic alterations of oncogenes and/or suppressor genes. In the past, endometrial cancer is classified as type I and type II according to hormone dependency for its pathogenesis [4]. Type I tumors are largely dependent on unopposed estrogen which is not suppressed by progestin, resulting in well-differentiated endometrioid tumors, and show better prognosis. Type I tumors arise from normal endometrium through atypical endometrial hyperplasia. Genetic alterations have been reported to be related to type I tumors, including mismatch repair genes (*MLH1*, *MSH2*, *MSH6*, *PMS2*), *PTEN*, and *PIK3CA*. Type II tumors are independent of estrogen and arise from atrophic endometrium through endometrial intraepithelial carcinoma, resulting in aggressive tumors, and show poorer prognosis. *TP53* gene alteration has been implicated in pathogenesis of type II tumors. However, The Cancer Genome Atlas (TCGA) reported that endometrial cancer can be classified into four molecular subtypes, including polymerase ε (*POLE*) mutant type, microsatellite instability-high (MSI-high) type, copy number-low type, and copy number-high type in 2013 [5], and the patients' prognosis can be stratified according to subtypes. Recent topics on the molecular mechanism of endometrial carcinogenesis are provided by Drs. Kotaro Takahashi and Kosuke Yoshihara from Niigata University.

1.3 Molecular Classification of Endometrial Cancer

As described above, endometrial cancer has been classified as type I and type II from clinicopathological and molecular pathological aspects. However, TCGA reported that endometrial cancer can be classified into four molecular subtypes [5]. Because these molecular subtypes have been reported to predict patients' survival in the literature, the latest WHO classification for female genital tumors firstly described TCGA molecular classification for endometrial cancer [6]. The International Federation of Gynecology and Obstetrics (FIGO) proposed a new staging system concerning molecular subtypes for endometrial cancer (FIGO 2023) [7], which is extremely complex compared with that of FIGO 2008. Regarding prognosis, patients with the *POLE* mutant type show excellent prognosis, and those with copy number-high type shows the worst prognosis. Both of MSI-high type and copy number-low type show intermediate prognosis [5]. Alternative classification methods using immunohistochemistry have been proposed, including ProMis E [8] and trans PORTEC [9], because TCGA classification using next-generation sequencing (NGS) could not be directly implemented in clinical practice in terms of the costs. In alternative methods, MSI-high and copy number-high types are replaced with mismatch repair protein (MLH1, MSH2, MSH6, PMS2) expression deficiency and p53 overexpression (2+) and/or p53 null expression, respectively. However, even in those alternative methods, sequencing for *POLE* mutation using genetic testing (Sanger method, polymerase chain reaction, NGS) is indispensable, but the standard method is not established and not covered by insurance at this moment in Japan. Currently, the Japanese Society of Pathology (JSP), Japan Society of Obstetrics and Gynecology (JSOG), and Japan Society of Gynecologic Oncology (JSGO) are discussing how to implement the FIGO 2023 staging system for endometrial cancer into clinical practice in Japan. Drs. Hiroyuki Yamazaki and Hiroshi Asano from Hokkaido University describe this issue.

1.4 Initial Surgery for Endometrial Cancer [2]

Surgery is the first choice of treatment for endometrial cancer, partly because of less sensitivity to radiotherapy compared with cervical cancer and probably less sensitivity to chemotherapy compared to ovarian cancer. Additionally, surgical staging has been adopted by FIGO and the Japan Society of Obstetrics and Gynecology (JSOG).

Regarding hysterectomy, total hysterectomy (extrafascial) or extended simple hysterectomy is recommended for patients presumed to have no vaginal or parametrial involvement. Modified radical hysterectomy or radical hysterectomy is suggested for patients with presumed vaginal or parametrial involvement for complete resection. In principle, ovarian preservation is not recommended, but it is suggested for young patients with endometrioid G1 and preoperative stage IA disease after fully explaining the risks for patients.

For preoperative stage IV patients, surgical intervention is suggested if total hysterectomy plus maximal tumor reduction is feasible.

Minimally invasive surgery is recommended for patients presumed to be in stage I and suggested for those presumed to be in stage II, but not in patients with advanced stage disease so far in Japan. Dr. Yoshito Terai from Kobe University describes this issue.

1.5 Lymphadenectomy in Endometrial Cancer

FIGO has recommended the surgical staging system, including systematic lymphadenectomy for endometrial cancer. Since lymph node metastasis is one of the most important poor prognostic factors for endometrial cancer, the diagnostic significance of lymphadenectomy has been established. Pathological diagnosis of lymph node metastasis is beneficial to identify those who need adjuvant therapy. As both pelvic and para-aortic lymph nodes are regional lymph nodes for endometrial cancer, we must surgically evaluate all regional lymph nodes to accurately identify nodal disease. However, lymphadenectomy is apparently associated with more blood loss, longer operation time, and possible increase of perioperative complications such as lymphedema and lymph cyst formation. If a higher extent of lymphadenectomy up to the level of the para-aortic region would result in a better prognosis, systematic lymphadenectomy up to the level of the renal vein would be strongly recommended for patients with preoperative risk of lymph node metastasis, but its survival benefit remains controversial. Currently, because two randomized controlled trials, AGO-Op.6 (ECLAT) in Germany [10] and Japan Clinical Oncology Group Study 1412 (JCOG1412) in Japan [11], are ongoing worldwide to evaluate the survival benefit of lymphadenectomy, we will get the results from the two trials in the near future. In the latest Japanese guideline [2], omission of pelvic lymphadenectomy is suggested in patients with endometrioid G1 or G2 that is presumed to be stage IA preoperatively. For patients who are preoperatively estimated to be at an intermediate or high risk of recurrence, pelvic lymphadenectomy is recommended, but in addition to pelvic lymphadenectomy, para-aortic lymphadenectomy down to the renal vein is suggested. Dr. Yosuke Konno from Hokkaido University discusses this issue.

Sentinel lymph node biopsy has been widely employed instead of extensive lymphadenectomy in endometrial cancer surgery to identify nodal disease and to avoid complications by lymphadenectomy in western countries as well as in certain Japanese institutions. In the latest Japanese guideline [2], omission of pelvic lymph node dissection by sentinel node biopsy is suggested as a clinical trial in medical institutes that are skilled in the technique of sentinel node biopsy and have the cooperation of pathologists for intraoperative diagnosis, because sentinel node biopsy has not been covered by medical insurance in Japan at this moment. Dr. Hideaki Yahata from Kyushu University describes this issue.

1.6 Drug Treatment for Endometrial Cancer

As mentioned in the introduction, the Japanese guideline [2] has recommended systemic chemotherapy, not radiotherapy, as adjuvant treatment for endometrial cancer (Fig. 1.2), because we believe that extensive surgery including systematic pelvic lymphadenectomy can reduce the risk of local recurrence, even without adjuvant radiotherapy, and we aim to reduce the risk of distant failure by adjuvant chemotherapy. Indeed, the Japanese Gynecologic Oncology Group study 2043 (JGOG 2043) post hoc analysis recently revealed that pelvic recurrence is reduced by more extensive pelvic lymphadenectomy (more than 20 pelvic nodes removed) [12].

The Japanese guideline [2] would recommend adriamycin and cisplatin (AP) as the standard chemotherapeutic regimen as adjuvant therapy for endometrial cancer based on the result of JGOG 2043 [13], and the paclitaxel and carboplatin (TC) regimen is suggested as well. For advanced and recurrent disease that are unresectable or have residual tumor, TC has been recognized as the standard regimen through the result of GOG 209 [14], showing the noninferiority of TC to paclitaxel, doxorubicin, and cisplatin (TAP). In recent years, immune checkpoint inhibitors (pembrolizumab, dostarlimab, durvalumab, atezolizumab) have been shown to be effective in combination with TC followed by maintenance with immune checkpoint inhibitors as the first-line treatment for advanced or recurrent cases [15–17]. Maintenance therapy using the durvalumab and olaparib combination is approved for MMR-proficient cases only [17]. For the second line treatment, the combination of multi-tyrosine kinase inhibitor targeting fibroblast growth factor receptor, vascular endothelial growth factor receptor, etc. (lenvatinib) with immune checkpoint inhibitor (pembrolizumab) has been approved [18]. Dr. Junzo Hamanishi from Kyoto University describes this issue.

Recently, a new class of drugs called ADC (antibody-drug conjugate) has been developed and clinically introduced in the treatment for a variety of solid tumors. ADC is composed of a monoclonal antibody covalently attached to a cytotoxic drug via a chemical linker. Four ADCs have been approved in Japan (trastumab emtansine, trastumab deruxtecan, sacituzumab govitecan, datopotamab deruxtecan), but only for breast cancer as of the end of December 2024. Efficacy and safety of ADCs have been tested for gynecologic cancers including endometrial cancer in clinical trials in western countries as well as in Japan. Trastumab deruxtecan (DS-08201a) is composed of a humanized anti-human HER2 (human epidermal growth factor receptor 2) antibody, an enzymatically cleavable peptide linker, and a exatecan-derivative topoisomerase 1 inhibitor. Sacituzumab govitecan (MK-2870) is composed of a humanized anti-human TROP-2 (trophoblast-cell surface antigen 2) antibody, an enzymatically cleavable peptide linker, and a berotecan-derivative topoisomerase 1 inhibitor. Both DS-08201a and MK-2870 are currently tested for endometrial cancer in clinical trials in Japan. Dr. Takashi Mitamura from Hokkaido University describes the details and latest information on this issue.

1.7 Fertility-Sparing Therapy

Younger patients with endometrial cancer are generally considered to be women under 40 years old. In Japan, the number of young patients has been gradually increasing due to the increase of the total number of patients with endometrial cancer, because the rate of young patients remains around 5% [1]. In the latest Japanese guideline [2], progestin therapy is suggested for patients with atypical endometrial hyperplasia or endometrioid carcinoma G1 limited to the endometrium, who strongly desire fertility preservation and understand the risk of disease recurrence. In addition, assisted reproductive technology is encouraged for conception. For patients with residual lesions or intrauterine recurrence after fertility-sparing treatment, conservative treatment is not routinely recommended for the patients with residual lesions, but repeat progestin therapy is suggested under close supervision in patients with intrauterine recurrence and a strong desire for fertility preservation in our guideline [2]. Currently, a clinical trial (JGOG2051) has been conducted to show the effectiveness and safety of repeat progestin treatment for recurrent cases in Japan [19]. Drs. Kensuke Sakai and Wataru Yamagami from Keio University describe this issue.

1.8 Endometrial Cancer as a Hereditary Disease

We, gynecologic oncologists, would encounter female patients at increased risk of breast, ovarian, endometrial, cervical, and colon cancers due to genetic predisposition. Hereditary cancer syndromes refer to a genetic predisposition to specific cancers due to inherited pathogenic mutations, characterized by early age onset, involvement of multiple organs, and occurrence among family members. Several hereditary cancer syndromes are known to be associated with endometrial cancer.

Lynch syndrome is characterized by mutations in DNA mismatch repair genes (*MLH1*, *MSH2*, *MSH6*, *PMS2*), which are highly associated with increased risk of colon cancer and endometrial cancer with estimated lifetime risks as high as 60% for both [20]. Other cancers associated with Lynch syndrome include ovarian, gastric, small bowel, hepatobiliary, kidney, and urethral cancers.

In Cowden syndrome, 20–34% of the patients are found to carry a germline *PTEN* mutation. Breast cancer is the most common malignancy associated with Cowden syndrome with a cumulative risk as high as 85%, but among female carriers, the risk of endometrial cancer is as high as 30% [20]. Other cancers associated with Cowden syndrome include follicular thyroid, renal cell, and colorectal cancers. Dr. Takashi Mitamura from Hokkaido University would describe this issue including surveillance and risk-reducing strategies for female carriers.

1.9 Uterine Sarcomas

Uterine sarcomas arise from mesenchymal tissues and include various histologic types. The most prevalent histologic type is leiomyosarcoma (261 cases) followed by low-grade endometrial stromal sarcoma (102 cases), high-grade endometrial stromal sarcoma (74 cases), and undifferentiated sarcoma (29 cases) among newly diagnosed uterine sarcoma cases according to the patient annual report for 2021 in Japan [1]. According to the latest Japanese guideline [2], total abdominal hysterectomy and bilateral salpingo-oophorectomy are recommended. Additionally, tumor reduction surgery with the aim of complete resection is suggested, but lymph node dissection is not suggested for leiomyosarcoma and low-grade endometrial stromal sarcoma without findings of lymph node metastasis. After complete surgical resection during initial surgery, post surgical adjuvant therapy is not suggested for stage I leiomyosarcoma, undifferentiated uterine sarcoma, and high-grade endometrial stromal sarcoma, but hormone therapy is suggested when postoperative adjuvant therapy is indicated for low-grade endometrial stromal sarcoma. For unresectable cases, chemotherapy (doxorubicin monotherapy) is recommended. Pazopanib, trabectedin, and eribulin are suggested in the second-line and subsequent treatments. Hormone therapy is also suggested for unresectable cases of low-grade endometrial stromal sarcoma.

Genetic analysis identified that 5–10% of uterine leiomyosarcoma cases show *BRCA2* somatic mutation and the effectiveness of PARP inhibitors has been reported as case series [21, 22]. Currently, a clinical trial (JGOG2052) [23] has been conducted to examine the efficacy and safety of niraparib for unresectable and recurrent leiomyosarcoma in Japan. Dr. Hiroshi Asano from Hokkaido University describes this issue.

1.10 Artificial Intelligence in the Diagnosis of Endometrial Cancer

Artificial intelligence (AI) refers to technologies and concepts that enable computers to mimic intelligent behavior. As AI progressed, it evolved into machine learning, focusing on building models that learn from data to make predictions. Deep learning, a subset of machine learning, addresses the challenges of handling unstructural data by utilizing artificial neural networks. AI is transforming medical diagnostics, especially in endoscopy, imaging (X-ray, MRI, CT scans), and waveform analysis (ECGs). In MRI-based preoperative diagnosis of endometrial cancer, AI would enhance tumor detection and myometrial invasion assessment. In addition, AI-assisted pathology using hematoxilin-eosin-stained slides would improve histopathologic diagnosis and molecular classification of endometrial cancer, which would affect treatment decision. Drs. Yusuke Toyohara and Kenbun Sone from the University of Tokyo describe the latest knowledge on the AI-guided diagnosis of endometrial cancer.

References

1. Japan Society of Obstetrics and Gynecology, Gynecologic Oncology Committee. Patients annual report for 2021 in Japan. 2023; pp. 1668.
2. Morisada T, Kobayashi Y, Nagase S, Baba T, Tokunaga H, Watari H, Takekuma M, Terao Y, Terai Y, Kajiyama H, Asai-Sato M, Takehara K, Tabata T, Harano K, Hirashima Y, Yunokawa M, Okamoto A, Mikami M. Japan Society of Gynecologic Oncology 2023 guidelines for treatment of uterine body neoplasm. J Gynecol Oncol. 2025;36(3):e119.
3. Concin N, Matias-Guiu X, Vergote I, Cibula D, Mirza MR, Marnitz S, Ledermann J, Bosse T, Chargari C, Fagotti A, Fotopoulou C, Martin AG, Lax S, Lorusso D, Marth C, Morice P, Nout RA, O'Donnell D, Querleu D, Raspollini MR, Sehouli J, Sturdza A, Taylor A, Westermann A, Wimberger P, Colombo N, Planchamp F, Creutzberg CL. ESGO/ESTRO/ESP guidelines for the management of patients with endometrial carcinoma. Radiother Oncol. 2021;154:327–53.
4. Bokhman JV. Two pathogenetic types of endometrial carcinoma. Gynecol Oncol. 1983;15(1):10–7.
5. Cancer Genome Atlas Research Network, Kandoth C, Schultz N, Cherniack AD, Akbani R, Liu Y, Shen H, Robertson AG, Pashtan I, Shen R, Benz CC, Yau C, Laird PW, Ding L, Zhang W, Mills GB, Kucherlapati R, Mardis ER, Levine DA. Integrated genomic characterization of endometrial carcinoma. Nature. 2013;497(7447):67–73.
6. Tumours of the uterine corpus. In: Female genital tumours, WHO classification of tumours, vol. 4. 5th ed. IARC; 2020. p. 256.
7. Berek JS, Matias-Guiu X, Creutzberg C, Fotopoulou C, Gaffney D, Kehoe S, Lindemann K, Mutch D, Concin N. Endometrial cancer staging subcommittee, FIGO women's cancer committee. J Gynecol Oncol. 2023;34(5):e85.
8. Talhouk A, McConechy MK, Leung S, Yang W, Lum A, Senz J, Boyd N, Pike J, Anglesio M, Kwon JS, Karnezis AN, Huntsman DG, Gilks CB, McAlpine JN. Confirmation of ProMisE: a simple, genomics-based clinical classifier for endometrial cancer. Cancer. 2017;123(5):802–13.
9. Stelloo E, Nout RA, Osse EM, Jürgenliemk-Schulz IJ, Jobsen JJ, Lutgens LC, van der Steen-Banasik EM, Nijman HW, Putter H, Bosse T, Creutzberg CL, Smit VT. Improved risk assessment by integrating molecular and clinicopathological factors in early-stage endometrial cancer-combined analysis of the PORTEC cohorts. Clin Cancer Res. 2016;22(16):4215–24.
10. Emons G, Kim JW, Weide K, de Gregorio N, Wimberger P, Trillsch F, Gabriel B, Denschlag D, Kommoss S, Aydogdu M, Papathemelis T, Gropp-Meier M, Muallem MZ, Kühn C, Müller A, Frank M, Weigel M, Bronger H, Lampe B, Rau J, Schade-Brittinger C, Harter P. Endometrial cancer lymphadenectomy trial (ECLAT) (ECLAT) (pelvic and para-aortic lymphadenectomy in patients with stage I and II endometrial cancer with high risk of recurrence; AGO-OP.6). Int J Gynecol Cancer. 2021;31(7):1075–9.
11. Watari H, Katayama H, Shibata T, Ushijima K, Satoh T, Onda T, Aoki D, Fukuda H, Yaegashi N, Sakuragi N. Gynecologic cancer study group of the Japan clinical oncology group. Phase III trial to confirm the superiority of pelvic and para-aortic lymphadenectomy to pelvic lymphadenetomy alone for endometrial cancer: Japan Clinical Oncology Group Study 1412 (SEPAL-P3). Jpn J Clin Oncol. 2017;47(10):986–90.
12. Konno Y, Mayama M, Takehara K, Yokoyama Y, Suzuki J, Susumu N, Harano K, Nakagawa S, Nakanishi T, Yamagami W, Yoshihara K, Nomura H, Okamoto A, Aoki D, Watari H. Prognostic impact of the number of resected pelvic nodes un endometrial cancer: Japanese Gynecologic Oncology Group Study JGOG2043 poat hoc analysis. J Gynecol Oncol. 2024;2025(36):e3.
13. Nomura H, Aoki D, Michimae H, Mizuno M, Nakai H, Arai M, Sasagawa M, Ushijima K, Sugiyama T, Saito M, Tokunaga H, Matoda M, Nakanishi T, Watanabe Y, Takahashi F, Saito T, Yaegashi N, Japanese Gynecologic Oncology Group. Effect of taxane plus platinum regimens vs doxorubicin plus cisplatin as adjuvant chemotherapy for endometrial cancer at a high risk of progression: a randomixed clinical trial. JAMA Oncol. 2019;5(6):833–40.
14. Miller DS, Filiaci VL, Mannel RS, Cohn DE, Matsumoto T, Tewari KS, DiSilvestro P, Pearl ML, Argenta PA, Powell MA, Zweizig SL, Warshal DP, Hanjani P, Carney ME, Huang H,

Cella D, Zaino R, Fleming GF. Carboplatin and paclitaxel for advanced endometrial cancer: final overall survival and adverse event analysis of a phase III trial (NRG oncology/GOG0209). J Clin Oncol. 2020;38(33):3841–50.
15. Eskander RN, Sill MW, Beffa L, Moore RG, Hope JM, Musa FB, Mannel R, Shahin MS, Cantuaria GH, Girda E, Mathews C, Kavecansky J, Leath CA 3rd, Gien LT, Hinchcliff EM, Lele SB, Landrum LM, Backes F, O'Cearbhaill RE, Al Baghdadi T, Hill EK, Thaker PH, John VS, Welch S, Fader AN, Powell MA, Aghajanian C. Pembrolizumab plus chemotherapy in advanced endometrial cancer. N Engl J Med. 2023;388(23):2159–70.
16. Mirza MR, Chase DM, Slomovitz BM, dePont Christensen R, Novák Z, Black D, Gilbert L, Sharma S, Valabrega G, Landrum LM, Hanker LC, Stuckey A, Boere I, Gold MA, Auranen A, Pothuri B, Cibula D, McCourt C, Raspagliesi F, Shahin MS, Gill SE, Monk BJ, Buscema J, Herzog TJ, Copeland LJ, Tian M, He Z, Stevens S, Zografos E, Coleman RL, Powell MA; RUBY Investigators. Dostarlimab for primary advanced or recurrent endometrial cancer. N Engl J Med 2023; 388 (23): 2145–2158.
17. Westin SN, Moore K, Chon HS, Lee JY, Thomes Pepin J, Sundborg M, Shai A, de la Garza J, Nishio S, Gold MA, Wang K, McIntyre K, Tillmanns TD, Blank SV, Liu JH, McCollum M, Contreras Mejia F, Nishikawa T, Pennington K, Novak Z, De Melo AC, Sehouli J, Klasa-Mazurkiewicz D, Papadimitriou C, Gil-Martin M, Brasiuniene B, Donnelly C, Del Rosario PM, Liu X, Van Nieuwenhuysen E, DUO-E Investigators. Durvalumab plus carboplatin/paclitaxel followed by maintainance durvalumab with or without olaparib as first-line treatment for advanced endometrial cancer: the phase III DUO-E rial. J Clin Oncol. 2024;42(3):283–99.
18. Makker V, Colombo N, Casado Herráez A, Santin AD, Colomba E, Miller DS, Fujiwara K, Pignata S, Baron-Hay S, Ray-Coquard I, Shapira-Frommer R, Ushijima K, Sakata J, Yonemori K, Kim YM, Guerra EM, Sanli UA, McCormack MM, Smith AD, Keefe S, Bird S, Dutta L, Orlowski RJ, Lorusso D, Study 309–KEYNOTE-775 Investigators. Lenvatinib plus pembrolizumab for advanced endometrial cancer. N Engl J Med. 2022;386(5):437–48.
19. Sakai K, Yamagami W, Sato Y, Susumu N, Yokoyama Y, Takehara K, Mandai M, Okamoto A. Phase II trial evaluating the efficacy and safety of repeated high dose medroxyprogesterone acetate (MPA) therapy for patients with recurrent early-stage endometrial cancer or atypical endometrial hyperplasia: Japanese Gynecologic Oncology Group study (JGOG2051/KGOG2031, REMPA trial). J Gynecol Oncol. 2024;35(6):e106.
20. González Peña T, Huang M. Genetic predisposition for gynecologic cancers. Clin Obstet Gynecol. 2024;67(4):660–4.
21. Hensley ML, Chavan SS, Solit DB, Murali R, Soslow R, Chiang S, Jungbluth AA, Bandlamudi C, Srinivasan P, Tap WD, Rosenbaum E, Taylor BS, Donoghue MTA, Hyman DM. Genomic landscape of uterine sarcomas defined through prospective clinical sequencing. Clin Cancer Res. 2020;26(14):3881–8.
22. Seligson ND, Kautto EA, Passen EN, Stets C, Toland AE, Millis SZ, Meyer CF, Hays JL, Chen JL. BRCA1/2 functional loss defines a targetable subset in leiomyosarcoma. Oncologist. 2019;24(7):973–9.
23. Asano H, Oda K, Yoshihara K, Ito YM, Matsumura N, Shimada M, Watari H, Enomoto T. Phase II study of niraparib in recurrent or persistent rare fraction of gynecologic malignancies with homologous recombination deficiency (JGOG2052). J Gynecol Oncol. 2022;33(4):e55.

Molecular Mechanism of Endometrial Carcinogenesis

2

Kotaro Takahashi and Kosuke Yoshihara

Abstract

The endometrium, which is the origin of endometrial cancer, already harbors cancer-associated gene mutations due to factors such as aging or menstruation, even though it is histologically normal. The complex plexus network of endometrial glands in the stratum basalis ("rhizome structure") contributes to the accumulation of genomic alterations, leading to clonal expansion of endometrial epithelial cells with cancer-associated gene mutations. Previous studies have identified driver genes such as *PTEN* and *TP53* in endometrial cancer and have shown that molecular subtype classifications reflect its biological characteristics more accurately than the conventional classification. Bridging the gap of genomic alterations between the endometrium and endometrial cancer is crucial for understanding the mechanisms of carcinogenesis. This chapter summarizes the genomic alterations and the associated molecular pathways in the endometrium and endometrial cancer and discusses the carcinogenesis of endometrial cancer. Further elucidation of these mechanisms is essential to enable the development of novel strategies for treating and preventing endometrial cancer.

Keywords

Endometrium · Endometrial cancer · Genomic alteration · Molecular subtype · Clonal evolution

K. Takahashi · K. Yoshihara (✉)
Department of Obstetrics and Gynecology, Niigata University Graduate School of Medical and Dental Sciences, Niigata, Japan
e-mail: yoshikou.med@niigata-u.ac.jp

H. Watari, T. Mitamura (eds.), *Uterine Body Neoplasms*, Comprehensive Gynecology and Obstetrics, https://doi.org/10.1007/978-981-95-0946-1_2

2.1 Introduction

Carcinogenesis of the endometrium is considered to involve cancer-associated gene mutations, such as *PTEN*, *PIK3CA*, *TP53*, *ARID1A*, and *KRAS* [1]. To detect true driver genes from cancer-associated genes, in vivo genetic models such as genetically engineered mouse models have been utilized [2]. However, recent advances in sequencing technologies have revealed that cancer-associated genes, including *PIK3CA* and *KRAS*, are highly mutated in the normal endometrium [3–5]. In addition, the endometrium structure in humans who menstruate differs significantly from that in mice who do not have a menstrual cycle [6]. These findings suggest that the mechanism of carcinogenesis in endometrial cancer is more complex than previously thought.

Endometrial cancer has conventionally been broadly classified into two categories: type I, which occurs around menopause and predominantly consists of endometrioid carcinoma, and type II, which occurs in older women and primarily includes non-endometrioid carcinoma exemplified by serous carcinoma [7]. However, in 2013, The Cancer Genome Atlas (TCGA) Research Network team proposed a classification system that divides endometrial cancer into four molecular subtypes, which have been shown to more accurately reflect the biological characteristics of this cancer than the conventional classification [8]. These molecular subtypes are also crucial for understanding the mechanisms of carcinogenesis.

This chapter will focus primarily on molecular features in the normal endometrium and endometrial cancer and discuss the molecular mechanisms of endometrial carcinogenesis.

2.2 Accumulation of Cancer-Associated Gene Mutations in the Endometrium

First, we focus on the endometrium, the inner lining of the uterus, where endometrial cancer arises. Before the advent of next-generation sequencing technology, it was thought that cancer-related genes in the normal endometrium were wild-type and not mutated. On the other hand, previous studies reported that histologically normal endometrium contains some glands without PTEN expression due to *PTEN* mutations [9, 10]. We used laser microdissection to selectively collect normal endometrial epithelial cells and performed whole-exome sequencing to evaluate the genomic alterations of normal endometrial glands. As a result, somatic mutations in cancer-associated genes, such as *PIK3CA*, *KRAS*, *FBXW7*, *PPP2R1A*, and *PIK3R1*, were identified in normal endometrial epithelium [3]. Recent studies have clarified that cancer-associated gene mutations exist not only in the normal endometrium but also in various normal tissues such as the skin, esophagus, colon, and bladder [11]. Furthermore, since it has been reported that the glandular structures, the smallest functional units of the endometrium, are monoclonal [12], we isolated individual glands from endometrial tissue and subjected them to genomic analysis. As expected, each gland harbors distinct somatic mutations and accumulates them in a monoclonal manner [3].

Next, two large-scale genomic analyses of the normal endometrium provided further insights into the characteristics of these mutations [4, 5]. These studies confirmed that more than half of the endometrial glands carry cancer-associated gene mutations. Among these, *PIK3CA* mutations were the most common, appearing in 15%–20% of the glands [4, 5]. Mutational signature analysis aimed at identifying the causes of the mutations showed a significant enrichment of patterns associated with aging. Indeed, the number of mutations was found to correlate with age [4, 5] and cumulative menstrual cycles [5].

To evaluate the significance of mutations in the endometrium, the dN/dS ratio (the ratio of nonsynonymous to synonymous substitutions), an indicator of the strength of natural selection, was analyzed. Intriguingly, cancer-associated genes in the endometrium were under positive selection. Notably, genes associated with gynecological cancers, such as *KRAS* and *PIK3CA*, were under particularly strong positive selection [4, 5]. These findings suggest that cancer-associated gene mutations in the endometrium may confer proliferative or survival advantages in the context of its cyclic regeneration with each menstrual cycle.

2.3 Clonal Expansion and Spatial Genomic Analysis of the Endometrium

Although endometrial glands are monoclonal, somatic mutations vary among individual glands [3]. Genomic analysis of endometrial glands, considering their spatial location within the tissue, demonstrated that glands with the same mutations were clustered together in specific regions [5]. For example, the diameter of these regions extended to 2 cm in the uterine endometrium of a 41-year-old woman with uterine myoma. To investigate the spatial expansion of identical clones in the endometrium, we performed a three-dimensional structural analysis using tissue-clearing techniques combined with light-sheet fluorescence microscopy [6]. This analysis revealed that endometrial glands form a network-like structure in the stratum basalis horizontally and extend vertically toward the stratum functionalis (Fig. 2.1). This horizontal structure was named the "rhizome" structure [6], and it is thought to contribute to regeneration of the endometrium because this structure is preserved during menstruation. Whole-genome sequencing analysis of individual glands connected by the rhizome structure demonstrated that these interconnected glands shared identical somatic mutations [5]. Thus, the rhizome structure, essential for postmenstrual regeneration of the endometrium, also leads to the accumulation of somatic mutations and the expansion of mutant clones (Fig. 2.1).

The timeline of genomic events can be estimated by analyzing whole-genome or whole-exome sequencing data of multiple glands within the same individual. In a 50-year-old case with uterine myoma, some clones exhibited copy-neutral loss of heterozygosity (CN-LOH) on chromosomes 3 and 10, along with a frameshift mutation in *PTEN* located on chromosome 10 [5]. Thus, *PTEN* had undergone two hits:

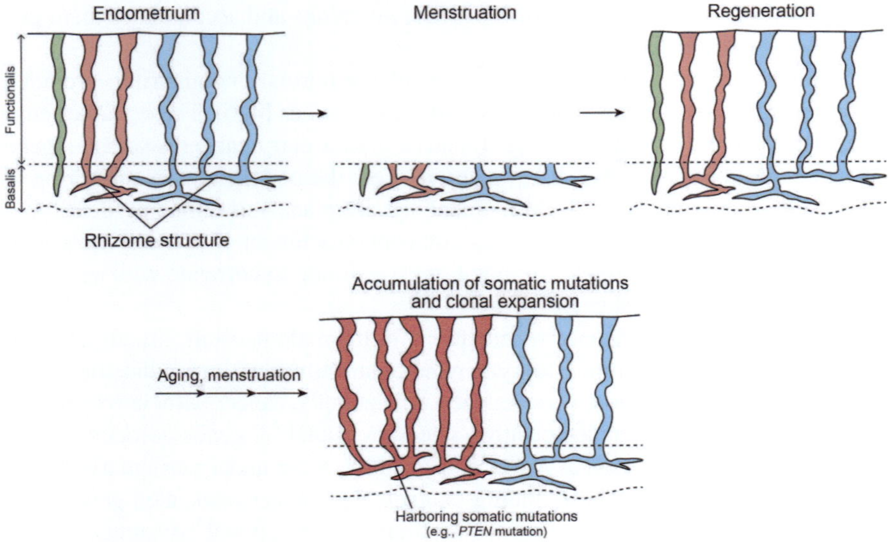

Fig. 2.1 Rhizome structure in the endometrium. The rhizome structure is maintained during menstruation and contributes to endometrial regeneration. Over time, genomic alterations accumulate and clonal expansion occurs in the endometrium, in which the rhizome structure is involved

a somatic mutation and copy number alterations. The molecular clock analysis suggested that the frameshift mutation in *PTEN* and the CN-LOH on chromosome 10 occurred in the patient's teens, and the CN-LOH on chromosome 3 occurred in her 30 s. This finding meant that decades had passed without cancer developing, even with these alterations [5]. In another study, it was similarly estimated that cancer-associated mutations can arise as early as the teenage years or even earlier [4]. A certain level of genomic alterations, including cancer-associated gene mutations, can be tolerated without developing cancer in the endometrium, probably due to tumor suppressor mechanisms [13].

2.4 Genomic Alterations and Molecular Subtypes Associated with Carcinogenesis in Endometrial Cancer

Next, we turn our attention to the genomic alterations in endometrial cancer. Conventionally, endometrial cancer has been broadly classified into two categories [7, 14]. Type I typically develops in middle-aged women around menopause, is predominantly of the endometrioid histological subtype, has a favorable prognosis, and is estrogen-dependent. It is often associated with atypical hyperplasia, which is considered a precancerous lesion. Meanwhile, type II arises in older women, is predominantly of the non-endometrioid histological subtypes (e.g., serous carcinoma, clear cell carcinoma), has a poor prognosis, and develops de novo from the atrophic endometrium.

A major paradigm shift in how endometrial cancer is characterized came in 2013 when TCGA proposed a molecular subtype classification for this disease [8]. On the basis of integrated genomic analysis, endometrial cancer was categorized into four molecular subtypes: *POLE* (ultramutated) (*POLE*), microsatellite instability (MSI) (hypermutated) (MSI), copy-number high (CN-H), and copy-number low (CN-L). These molecular subtypes were shown to have distinct prognoses: The *POLE* subtype exhibits an excellent prognosis, the CN-H subtype a poor prognosis, and the MSI and CN-L subtypes an intermediate prognosis. The outcome of the MSI subtype has significantly improved with the help of the immune checkpoint inhibitor. This molecular subtyping is beneficial not only for stratifying patients according to the most appropriate treatment but also for understanding the mechanism of carcinogenesis.

For clinical application, a method for classifying endometrial cancer using immunohistochemistry for mismatch repair (MMR) and p53 proteins has also been proposed. Subtypes classified by this method—*POLE*mut, MMR-deficient (MMRd), p53abn, and no specific molecular profile (NSMP)—correlate with prognosis, similar to the case for TCGA classification. WHO 2020 [15], ESGO/ESTRO/ESP [16], FIGO 2023 [17], and NCCN guidelines [18] also refer to this classification. In this chapter, we also use this clinical classification (*POLE*mut, MMRd, p53abn, and NSMP). When reading academic papers, the definition of the classification should be verified because the CN-H subtype based on genomic data may classify tumors not by p53 abnormalities but by copy number alterations [19, 20]. Clinical details of the molecular subtypes, including FIGO 2023, are discussed in Chap. 4. Here, we focus on aspects related to the mechanisms of carcinogenesis.

From a biological perspective, the prognosis of endometrial cancer can vary based on the molecular subtype, even within the same histological type. High-grade tumors, including endometrioid carcinoma grade 3 [21, 22], clear cell carcinoma [23], un/dedifferentiated carcinoma [24], and carcinosarcoma [24, 25], have been shown to exhibit distinct prognostic outcomes depending on their molecular subtype. This finding suggests that tumors with the same histological appearance exhibit different biological behaviors depending on their underlying genomic alterations. In mixed carcinomas, the two distinct histological types generally share common mutational clones, and some cases show additional mutations in distinct histological types [26, 27]. This implies that genomic alterations in the background play a key role in carcinogenesis and mutations that subsequently accumulate influence the histological type or phenotype. These findings underscore the importance of attempting to understand carcinogenesis through the lens of molecular subtypes. In this context, we provide the genomic and molecular biological perspectives for each subtype (Fig. 2.2).

2.4.1 *POLE*mut Subtype

POLE encodes the catalytic subunit of the replicative DNA polymerase epsilon, which contributes to DNA repair and chromosomal DNA replication. In the *POLE*mut subtype, missense mutations occur in the exonuclease domain of *POLE*,

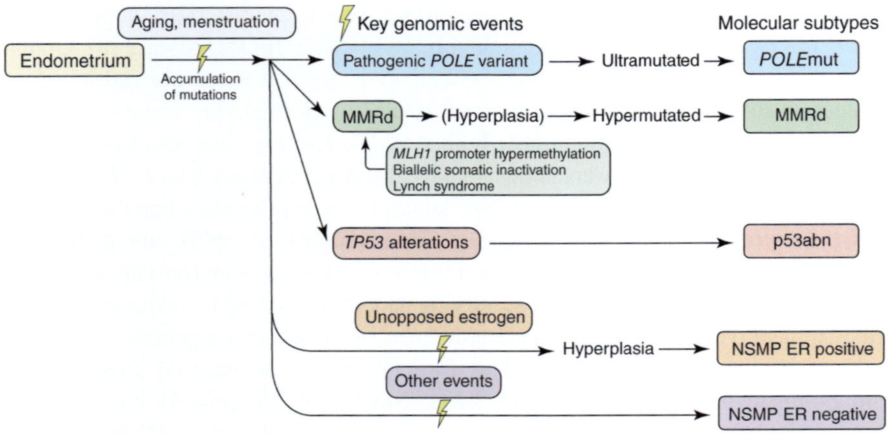

Fig. 2.2 Molecular mechanisms of carcinogenesis in the endometrium. This figure summarizes the genomic and molecular alterations assumed to occur in the progression from the noncancerous endometrium to each molecular subtype of endometrial cancer

resulting in impaired proofreading function and the abnormal accumulation of mutations. In many cases, tumor mutation burden (TMB) exceeds 100 mutations per megabase (mut/Mb), and thus, this subtype is also referred to as the ultramutated type [8, 28, 29]. The most common pathogenic variants are P286R and V411L, followed by S297F, A456P, and S459F. The mutations also accumulate with distinct patterns, such as C-to-A transversion accounting for over 20% of the total, while C-to-G transversion account for less than 0.6%. Pathogenic *POLE* variants can also be defined on the basis of the accumulation of mutations detected through whole-genome or whole-exome sequencing [28]. On the other hand, a clinical trial defines the *POLE*mut subtype as including a total of 11 pathogenic variants [30].

No comprehensive reports have indicated the presence of pathogenic *POLE* variants in the normal endometrium or atypical hyperplasia, and there has been no direct investigation into whether *POLE* mutation is an early event. Meanwhile, a disease called "polymerase proofreading-associated polyposis" (PPAP) has been identified, which is an autosomal dominant inherited disease that causes multiple colorectal adenomas and colorectal cancer due to germline mutations in *POLE* or *POLD1* [1, 31]. The existence of this disease suggests that pathogenic *POLE* variants cause mutations to accumulate, facilitating carcinogenesis.

The *POLE*mut subtype has been observed in various histological types. While it has not been reported in serous carcinoma or mesonephric-like carcinoma, it accounts for approximately 5%–12% of cases of low-grade endometrioid carcinoma, high-grade endometrioid carcinoma, clear cell carcinoma, mixed carcinoma, un/dedifferentiated carcinoma, and neuroendocrine tumors [29, 32]. Regardless of the histological type, the prognosis is remarkably favorable. This subtype clearly demonstrates that molecular subtypes are more critical than histological types in determining the biological behavior of cancer.

2.4.2 MMRd Subtype

Abnormalities in MMR genes, specifically *MLH1*, *MSH2*, *MSH6*, and *PMS2*, lead to DNA repair errors and the accumulation of mutations. Consequently, MSI is observed, and the TMB exceeds 10 mut/Mb, with this subtype also being referred to as the hypermutated type [8, 29]. The causes of MMRd are broadly classified into three categories: 72% result from hypermethylation of the *MLH1* promoter, 10% are associated with germline mutations in Lynch syndrome, and 18% are due to biallelic somatic inactivation of *MLH1*, *MSH2*, *MSH6*, *PMS2*, or *EPCAM* [33].

In endometrial cancer accompanied by atypical hyperplasia, some cases exhibit MMRd in atypical hyperplastic lesions [34, 35]. As a cause of this, *MLH1* promoter methylation has been observed in atypical hyperplasia [36, 37]. Furthermore, in Lynch syndrome cases, MMRd has also been reported in the normal endometrium [38]. These findings suggest that MMRd arises as an early event, and the accumulation of mutations based on MMR deficiency contributes to carcinogenesis.

2.4.3 p53abn Subtype

Abnormalities in p53 are associated with a high frequency of copy number alterations. Serous carcinoma is representative of the p53abn subtype. Atrophic endometrial epithelium with *TP53* mutations progresses to serous carcinoma via precancerous lesions of serous endometrial intraepithelial carcinoma [1, 39]. In fact, there are cases where normal endometrial tissue contains a small number of normal endometrial glands with *TP53* mutations [5]. Other histological types also show a high prevalence of p53 abnormalities in high-grade cases, with 73.9% of carcinosarcomas, 61.1% of mixed carcinomas, and 42.5% of clear cell carcinomas classified as p53abn [29, 32]. Regardless of histological types, the p53 pathway alteration, as discussed later, is thought to play a critical role in carcinogenesis in the p53abn subtype.

The mutations that coexist in cases of the p53abn subtype also differ from those in other molecular subtypes, with *FBXW7* and *PPP2R1A* mutations characteristically observed in approximately 20% of cases each [8, 40, 41]. Amplification of *ERBB2* and abnormal immunostaining of HER2, encoded by *ERBB2*, are also features of the p53abn subtype [42, 43]. HER2 positivity in endometrial cancer has been reported in approximately 10%–50% of cases [44], with over 90% of these classified as the p53abn subtype [45]. Clinically, the p53abn subtype has the poorest prognosis. This highlights the clinical and biological significance of the p53abn subtype as a distinct and aggressive form of endometrial cancer.

2.4.4 NSMP Subtype

The NSMP subtype represents a heterogeneous group that lacks the defining characteristics of the other three molecular subtypes. Unopposed estrogen stimulation is a significant risk factor for endometrial cancer [1], and it has been shown that estrogen-dependent tumors are generally classified into the NSMP subtype [46], with estrogen receptor (ER) positivity reported in 84% of NSMP cases [47]. Within the NSMP subtype, the prognosis clearly differs between ER-positive and ER-negative cases, with reports indicating that ER-negative cases in high-risk groups have a worse prognosis than those classified as the p53abn subtype [48]. This suggests that ER-positive and ER-negative NSMP cases exhibit distinct biological characteristics.

ER-positive patients tend to be relatively young and obese, to have low-grade endometrial cancer, and to exhibit the coexistence of atypical hyperplasia [47, 49]. While ER positivity is observed in other molecular subtypes [50], a study on fertility-preserving medroxyprogesterone acetate therapy reported that, even among ER-positive cases, those with the NSMP subtype had a complete response rate of 53.3%, compared with only 11% in the MMRd subtype [51]. This suggests that estrogen plays a crucial role in the carcinogenesis and maintenance of ER-positive NSMP tumors.

ER-negative cases within NSMP include high-grade tumors, such as clear cell carcinoma and mesonephric-like carcinoma. ER-negative cases are also characterized by a higher frequency of L1CAM immunohistochemical positivity, a known poor prognostic factor, compared with ER-positive cases [47–49]. As mentioned above, the NSMP subtype represents a heterogeneous group with no characteristic genomic alterations, suggesting the value of further efforts to subclassify this subtype or investigate its carcinogenic mechanisms through multi-omics analyses.

2.5 Molecular Pathways Associated with Endometrial Cancer

This section highlights the molecular pathways linked to the cancer-associated gene mutations frequently observed in endometrial cancer. Cancer-associated gene mutations lead to the activation or inactivation of specific molecular pathways, especially those involved in cellular proliferation or the determination of cell fate. Although the factors driving the accumulation of mutations may differ depending on the molecular subtype, 80%–95% of endometrial cancers exhibit mutations associated with the PI3K-PTEN-Akt-mTOR pathway [8, 42]. Therefore, understanding the related molecular pathways is crucial, regardless of the underlying causes of mutation accumulation. Below, we describe four representative molecular pathways with a focus on the cancer-associated gene mutations in endometrial cancer.

2.5.1 The PI3K-PTEN-Akt-mTOR Pathway

The PI3K-PTEN-AKT-mTOR pathway is a key molecular pathway involved in essential cellular processes, including proliferation, growth, and survival [42, 52, 53]. PI3K consists of the regulatory subunit PI3Kp85α, encoded by *PIK3R1*, and the catalytic subunit PI3Kp110α, encoded by *PIK3CA*. PI3K is activated by tyrosine kinase receptor stimulation via growth factor receptors such as EGFR and HER2, converting PIP2 into PIP3 at the cell membrane. PIP3 recruits PDK1, leading to the phosphorylation of AKT, which activates the mTOR pathway and promotes specific cellular processes, such as stimulating cell cycle progression and enhancing protein synthesis. PTEN, a tumor suppressor, inhibits this pathway by dephosphorylating PIP3 back to PIP2. As a result, loss of PTEN function leads to activation of this pathway (Fig. 2.3a). Additionally, ARID1A plays a regulatory role in suppressing this pathway. ARID1A also functions as part of the SWItch/Sucrose Non-Fermentable complex, contributing to chromatin remodeling and regulating gene expression [54]. Thus, loss of ARID1A function contributes to tumor progression.

PTEN, *PIK3CA*, *PIK3R1*, and *ARID1A* mutations are frequently observed in *POLE*mut, MMRd, and NSMP subtypes, with mutation frequencies of approximately 70%–90% for *PTEN*, 40%–80% for *PIK3CA*, 20%–60% for *PIK3R1*, and 40%–80% for *ARID1A* [8, 40, 41]. Notably, *PTEN* mutations are observed in

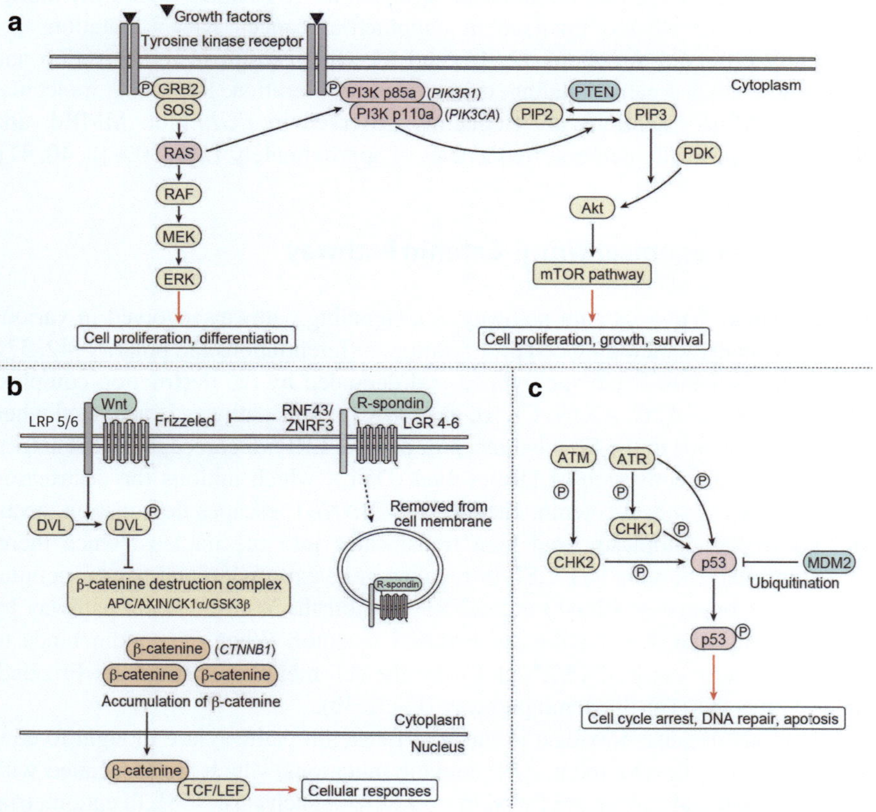

Fig. 2.3 Molecular pathways associated with endometrial cancer. (**a**) The PI3K-PTEN-Akt-mTOR pathway and the RAS-RAF-MEK-ERK-MAPK pathway. (**b**) The canonical Wnt-β-catenin pathway. (**c**) The p53 pathway

77%–92% of cases with concurrent atypical hyperplasia [35, 55], suggesting that they occur as an early event. The allele frequency of *PTEN* mutation is increasing from the normal endometrium to endometrial cancer via atypical hyperplasia [56]. This pathway is an essential molecular mechanism in many cases of endometrial cancer.

2.5.2 The RAS-RAF-MEK-ERK-MAPK Pathway

The RAS-RAF-MEK-ERK-MAPK pathway is a pivotal signaling cascade involved in cell proliferation and differentiation [42, 53]. It also plays an essential role in the growth of normal cells. Upon stimulation of tyrosine kinase receptors, GRB2 and SOS activate RAS, leading to the sequential phosphorylation of RAF, MEK, and ERK, which ultimately promotes cell proliferation. RAS also contributes to the activation of the PI3K pathway described above (Fig. 2.3a). In cancers, genetic or molecular alterations in RAS, RAF, and ERK generate strong signaling that promotes abnormal cell proliferation. KRAS is one of the RAS family members (alongside HRAS and NRAS) involved in endometrial cancer. *KRAS* mutations are frequently observed at codons 12, 13, and 61, which result in its activation and constitutive downstream signaling, driving cell proliferation. Regarding molecular subtypes, *KRAS* mutations are frequently observed in *POLE*mut, MMRd, and NSMP subtypes with mutation frequencies of approximately 15%–30% [8, 40, 41].

2.5.3 The Canonical Wnt-β-Catenin Pathway

The canonical Wnt-β-catenin pathway is a signaling pathway involved in various cellular responses, including cell proliferation, differentiation, and polarity [42, 57]. Normally, β-catenin is phosphorylated and degraded by the destruction complex, which consists of APC, AXIN, CK1α, and GSK3β. This pathway is activated when a Wnt ligand binds to the Frizzled receptor and the LRP5/6 coreceptor. This activation leads to the activation of Dishevelled (DVL), which inhibits the destruction complex. As a result, β-catenin, encoded by *CTNNB1*, escapes degradation, accumulates in the cytoplasm, and then translocates into the nucleus. Once there, β-catenin interacts with TCF/LEF to regulate gene expression and trigger cellular responses. Conversely, RNF43 and ZNRF3 inhibit the Wnt-β-catenin pathway by ubiquitinating and deactivating the Frizzled receptor. When R-spondin binds to LGR4–6, it removes RNF43/ZNRF3 from the cell membrane, stabilizes Frizzled, and activates the Wnt-β-catenin pathway (Fig. 2.3b).

Mutations in genes involved in the Wnt-β-catenin pathway are thought to contribute to cancer development. *APC* somatic mutations, which are associated with sporadic colorectal cancer, are found in over 80% of such cases [58]. In endometrial cancer, *CTNNB1* mutations are frequently observed in the NSMP subtype, at a rate of approximately 35%–50% [8, 40]. Notably, *CTNNB1* and *KRAS* mutations are mutually exclusive, suggesting that the two pathways may interact or that either mutation alone is sufficient for tumor formation [59]. Although less common than the previously mentioned mutations, *RNF43* mutations are reported in 18%–27% of endometrioid carcinomas [42, 60] and are similarly associated with the activation of the Wnt-β-catenin pathway.

2.5.4 The p53 Pathway

The p53 pathway is a critical signaling pathway that maintains genomic stability by orchestrating cellular responses to DNA damage and stress [61, 62]. Normally, p53 is ubiquitinated and degraded by MDM2, maintaining it at low levels. However, when cells are exposed to stress such as DNA damage, ATM, ATR, and CHK1/2 phosphorylate p53, converting it into its activated form. Activated p53 escapes degradation by MDM2 and promotes the transcription of p53 target genes. This leads to cellular responses such as cell cycle arrest, DNA repair, and apoptosis (Fig. 2.3c).

Alterations in *TP53* are closely associated with distinct p53 immunohistochemical patterns [63, 64]. Missense mutations in *TP53* result in nonfunctional mutant p53. Mutant p53 is resistant to degradation and accumulates in the nucleus, and immunohistochemistry exhibits an overexpression pattern. Conversely, when p53 function is lost due to truncating mutations in *TP53*, p53 activity is abrogated, and immunohistochemistry shows a null pattern with no staining. Additionally, when a truncating mutation occurs in the nuclear localization domain of *TP53*, p53 cannot translocate from the cytoplasm to the nucleus, resulting in the loss of its normal function and its accumulation in the cytoplasm. In such cases, immunohistochemistry shows a cytoplasmic pattern. The p53abn subtype is defined based on these abnormal p53 immunohistochemical patterns, reflecting alterations in the p53 pathway.

2.6 Clonal Evolution in Endometrial Cancer

Up to this point, we have discussed genomic alterations in the endometrium and endometrial cancer. It has been proposed that carcinogenesis progresses through the sequential involvement of cancer-associated genes corresponding to the breakthrough, expansion, and invasive phases [65]. A mathematical model analyzing genomic data of lung and colorectal cancers suggested that three driver gene mutations may be sufficient for carcinogenesis [66]. However, in endometrial cancer, the specific mutations and their number required for carcinogenesis have not been fully elucidated, and thus, these issues warrant further investigation.

After carcinogenesis, mutations can continue to accumulate, allowing the cancer to evolve. During this process, subclones with invasive or metastatic potential emerge, contributing to cancer progression. These subclones clonally expand at sites of invasion or metastasis [67, 68]. In endometrial cancer, multiple sampling of primary and metastatic lesions has shown that some subclones originating from the primary tumor metastasize and further accumulate mutations during their progression [55, 69, 70]. In our study that analyzed an endometrial cancer case of Lynch syndrome, mutational signature analysis of subclones showed that all subclones exhibited an increased signature associated with MMRd. This finding indicates that, even after cancer development, MMRd provided the basis for the continuous accumulation of mutations [70]. Furthermore, in an analysis comparing molecular subtypes between primary and recurrent cancer, 22 out of 24 cases maintained the same molecular subtype [71]. In the remaining two cases, the primary tumors were classified as NSMP, while the recurrent tumors were classified as MMRd, which was acquired during progression due to *MLH1* hypermethylation. Thus, the alterations driving mutational accumulation, including genomic and epigenetic alterations, play a crucial role in cancer development and evolution.

2.7 Conclusion

When genomic alterations occur in the endometrium, the rhizome structure contributes to their accumulation and clonal expansion. While such genomic alterations may confer survival advantages on the endometrium, they could also contribute to carcinogenesis. In endometrial cancer, characteristic genomic alterations are used to define specific molecular subtypes. However, it remains unclear how many or what types of alterations need to accumulate in the endometrium to trigger carcinogenesis. To elucidate the mechanisms of carcinogenesis, it is essential to bridge the gap between the genomic alterations in the endometrium and those in endometrial cancer. To this end, it will be important to track the expansion of mutant clones through three-dimensional structural analysis, multiple sampling, including at metastatic sites, and longitudinal sampling. Methods using cultured resources, such as endometrial organoids, would also be beneficial. It is anticipated that the integration of multi-omics analyses with data obtained via these approaches will deepen our understanding of the carcinogenesis. Research into the mechanisms of carcinogenesis should also contribute to the development of new strategies for treating and preventing endometrial cancer.

Acknowledgments We thank Edanz (https://jp.edanz.com/ac) for editing a draft of this manuscript.

References

1. Urick ME, Bell DW. Clinical actionability of molecular targets in endometrial cancer. Nat Rev Cancer. 2019;19:510–21.
2. Friel AM, Growdon WB, McCann CK, Olawaiye AB, Munro EG, Schorge JO, et al. Mouse models of uterine corpus tumors: clinical significance and utility. Front Biosci (Elite Ed). 2010;2:882–905.
3. Suda K, Nakaoka H, Yoshihara K, Ishiguro T, Tamura R, Mori Y, et al. Clonal expansion and diversification of cancer-associated mutations in endometriosis and normal endometrium. Cell Rep. 2018;24:1777–89.

4. Moore L, Leongamornlert D, Coorens THH, Sanders MA, Ellis P, Dentro SC, et al. The mutational landscape of normal human endometrial epithelium. Nature. 2020;580:640–6.
5. Yamaguchi M, Nakaoka H, Suda K, Yoshihara K, Ishiguro T, Yachida N, et al. Spatiotemporal dynamics of clonal selection and diversification in normal endometrial epithelium. Nat Commun. 2022;13:943.
6. Yamaguchi M, Yoshihara K, Suda K, Nakaoka H, Yachida N, Ueda H, et al. Three-dimensional understanding of the morphological complexity of the human uterine endometrium. iScience. 2021;24:102258.
7. Hecht JL, Mutter GL. Molecular and pathologic aspects of endometrial carcinogenesis. J Clin Oncol. 2006;24:4783–91.
8. Kandoth C, Schultz N, Cherniack AD, Akbani R, Liu Y, Shen H, et al. Integrated genomic characterization of endometrial carcinoma. Nature. 2013;497:67–73.
9. Mutter GL, Ince TA, Baak JP, Kust GA, Zhou XP, Eng C. Molecular identification of latent precancers in histologically normal endometrium. Cancer Res. 2001;61:4311–4.
10. Mutter GL, Monte NM, Neuberg D, Ferenczy A, Eng C. Emergence, involution, and progression to carcinoma of mutant clones in normal endometrial tissues. Cancer Res. 2014;74:2796–802.
11. Kakiuchi N, Ogawa S. Clonal expansion in non-cancer tissues. Nat Rev Cancer. 2021;21:239–56.
12. Tanaka M, Kyo S, Kanaya T, Yatabe N, Nakamura M, Maida Y, et al. Evidence of the monoclonal composition of human endometrial epithelial glands and mosaic pattern of clonal distribution in luminal epithelium. Am J Pathol. 2003;163:295–301.
13. Kennedy SR, Zhang Y, Risques RA. Cancer-associated mutations but no cancer: insights into the early steps of carcinogenesis and implications for early cancer detection. Trends Cancer. 2019;5:531–40.
14. Liu FS. Molecular carcinogenesis of endometrial cancer. Taiwan J Obstet Gynecol. 2007;46:26–32.
15. WHO Classification of Tumours Editorial Board. Female genital tumours. 5th ed. Lyon: International Agency for Research on Cancer; 2020.
16. Concin N, Matias-Guiu X, Vergote I, Cibula D, Mirza MR, Marnitz S, et al. ESGO/ESTRO/ESP guidelines for the management of patients with endometrial carcinoma. Int J Gynecol Cancer. 2021;31:12–39.
17. Berek JS, Matias-Guiu X, Creutzberg C, Fotopoulou C, Gaffney D, Kehoe S, et al. FIGO staging of endometrial cancer: 2023. Int J Gynaecol Obstet. 2023;162:383–94.
18. National Comprehensive Cancer Network. NCCN guidelines uterine neoplasms version 1.2025. National Comprehensive Cancer Network, Inc.; 2024.
19. Dou Y, Kawaler EA, Cui Zhou D, Gritsenko MA, Huang C, Blumenberg L, et al. Proteogenomic characterization of endometrial carcinoma. Cell. 2020;180:729–48.e26.
20. Dou Y, Katsnelson L, Gritsenko MA, Hu Y, Reva B, Hong R, et al. Proteogenomic insights suggest druggable pathways in endometrial carcinoma. Cancer Cell. 2023;41:1586–605.e15.
21. Bosse T, Nout RA, McAlpine JN, McConechy MK, Britton H, Hussein YR, et al. Molecular classification of grade 3 endometrioid endometrial cancers identifies distinct prognostic subgroups. Am J Surg Pathol. 2018;42:561–8.
22. Cun HT, Bernard L, Lande KT, Lawson BC, Nesbakken AJ, Davidson B, et al. Comprehensive molecular characterization of early stage grade 3 endometrioid endometrial adenocarcinoma. Gynecol Oncol. 2024;189:138–45.
23. Travaglino A, Raffone A, Santoro A, Raimondo D, Angelico G, Valente M, et al. Clear cell endometrial carcinomas with mismatch repair deficiency have a favorable prognosis: a systematic review and meta-analysis. Gynecol Oncol. 2021;162:804–8.
24. Hammer PM, Wang A, Vermij L, Zdravkovic S, Heilbroner L, Ryan E, et al. Molecular classification outperforms histologic classification in prognostication of high-grade endometrial carcinomas with undifferentiated and sarcomatous components. Am J Surg Pathol. 2024;48:953–64.
25. Gotoh O, Sugiyama Y, Takazawa Y, Kato K, Tanaka N, Omatsu K, et al. Clinically relevant molecular subtypes and genomic alteration-independent differentiation in gynecologic carcinosarcoma. Nat Commun. 2019;10:4965.

26. Köbel M, Meng B, Hoang LN, Almadani N, Li X, Soslow RA, et al. Molecular analysis of mixed endometrial carcinomas shows clonality in most cases. Am J Surg Pathol. 2016;40:166–80.
27. Matrai C, Motanagh S, Mirabelli S, Ma L, He B, Chapman-Davis E, et al. Molecular profiles of mixed endometrial carcinoma. Am J Surg Pathol. 2020;44:1104–11.
28. León-Castillo A, Britton H, McConechy MK, McAlpine JN, Nout R, Kommoss S, et al. Interpretation of somatic POLE mutations in endometrial carcinoma. J Pathol. 2020;250:323–35.
29. How JA, Jazaeri AA, Westin SN, Lawson BC, Klopp AH, Soliman PT, et al. Translating biological insights into improved management of endometrial cancer. Nat Rev Clin Oncol. 2024;21:781–800.
30. Consortium RR. Refining adjuvant treatment in endometrial cancer based on molecular features: the RAINBO clinical trial program. Int J Gynecol Cancer. 2022;33:109–17.
31. Mur P, García-Mulero S, Del Valle J, Magraner-Pardo L, Vidal A, Pineda M, et al. Role of POLE and POLD1 in familial cancer. Genet Med. 2020;22:2089–100.
32. Arciuolo D, Travaglino A, Raffone A, Raimondo D, Santoro A, Russo D, et al. TCGA molecular prognostic groups of endometrial carcinoma: current knowledge and future perspectives. Int J Mol Sci. 2022;23:11684.
33. Post CCB, Stelloo E, Smit VTHB, Ruano D, Tops CM, Vermij L, et al. Prevalence and prognosis of lynch syndrome and sporadic mismatch repair deficiency in endometrial cancer. J Natl Cancer Inst. 2021;113:1212–20.
34. Russo M, Broach J, Sheldon K, Houser KR, Liu DJ, Kesterson J, et al. Clonal evolution in paired endometrial intraepithelial neoplasia/atypical hyperplasia and endometrioid adenocarcinoma. Hum Pathol. 2017;67:69–77.
35. Chapel DB, Patil SA, Plagov A, Puranik R, Mendybaeva A, Steinhardt G, et al. Quantitative next-generation sequencing-based analysis indicates progressive accumulation of microsatellite instability between atypical hyperplasia/endometrial intraepithelial neoplasia and paired endometrioid endometrial carcinoma. Mod Pathol. 2019;32:1508–20.
36. Horowitz N, Pinto K, Mutch DG, Herzog TJ, Rader JS, Gibb R, et al. Microsatellite instability, MLH1 promoter methylation, and loss of mismatch repair in endometrial cancer and concomitant atypical hyperplasia. Gynecol Oncol. 2002;86:62–8.
37. Kanaya T, Kyo S, Sakaguchi J, Maida Y, Nakamura M, Takakura M, et al. Association of mismatch repair deficiency with PTEN frameshift mutations in endometrial cancers and the precursors in a Japanese population. Am J Clin Pathol. 2005;124:89–96.
38. Hegazy S, Brand RE, Dudley B, Karloski E, Lesnock JL, Elishaev E, et al. Mutation-specific mismatch repair-deficient benign endometrial glands in endometrial biopsies and curettings are a biomarker of lynch syndrome and associate with endometrial carcinoma development. Am J Surg Pathol. 2023;47:835–43.
39. Zheng W, Xiang L, Fadare O, Kong B. A proposed model for endometrial serous carcinogenesis. Am J Surg Pathol. 2011;35:e1–e14.
40. Rios-Doria E, Momeni-Boroujeni A, Friedman CF, Selenica P, Zhou Q, Wu M, et al. Integration of clinical sequencing and immunohistochemistry for the molecular classification of endometrial carcinoma. Gynecol Oncol. 2023;174:262–72.
41. Asami Y, Kobayashi Kato M, Hiranuma K, Matsuda M, Shimada Y, Ishikawa M, et al. Utility of molecular subtypes and genetic alterations for evaluating clinical outcomes in 1029 patients with endometrial cancer. Br J Cancer. 2023;128:1582–91.
42. Bell DW, Ellenson LH. Molecular genetics of endometrial carcinoma. Annu Rev Pathol. 2019;14:339–67.
43. Ross DS, Devereaux KA, Jin C, Lin DY, Zhang Y, Marra A, et al. Histopathologic features and molecular genetic landscape of HER2-amplified endometrial carcinomas. Mod Pathol. 2022;35:962–71.
44. Halle MK, Tangen IL, Berg HF, Hoivik EA, Mauland KK, Kusonmano K, et al. HER2 expression patterns in paired primary and metastatic endometrial cancer lesions. Br J Cancer. 2018;118:378–87.

45. Vermij L, Horeweg N, Leon-Castillo A, Rutten TA, Mileshkin LR, Mackay HJ, et al. HER2 status in high-risk endometrial cancers (PORTEC-3): relationship with histotype, molecular classification, and clinical outcomes. Cancers (Basel). 2020;13:44.
46. Sugiyama Y, Gotoh O, Fukui N, Tanaka N, Hasumi K, Takazawa Y, et al. Two distinct tumorigenic processes in endometrial endometrioid adenocarcinoma. Am J Pathol. 2020;190:234–51.
47. Jamieson A, Huvila J, Chiu D, Thompson EF, Scott S, Salvador S, et al. Grade and estrogen receptor expression identify a subset of no specific molecular profile endometrial carcinomas at a very low risk of disease-specific death. Mod Pathol. 2023;36:100085.
48. Vermij L, Jobsen JJ, León-Castillo A, Brinkhuis M, Roothaan S, Powell ME, et al. Prognostic refinement of NSMP high-risk endometrial cancers using oestrogen receptor immunohistochemistry. Br J Cancer. 2023;128:1360–8.
49. Momeni-Boroujeni A, Nguyen B, Vanderbilt CM, Ladanyi M, Abu-Rustum NR, Aghajanian C, et al. Genomic landscape of endometrial carcinomas of no specific molecular profile. Mod Pathol. 2022;35:1269–78.
50. Vrede SW, Van Weelden WJ, Bulten J, Gilks CB, Teerenstra S, Huvila J, et al. Hormonal biomarkers remain prognostically relevant within the molecular subgroups in endometrial cancer. Gynecol Oncol. 2024;192:15–23.
51. Chung YS, Woo HY, Lee JY, Park E, Nam EJ, Kim S, et al. Mismatch repair status influences response to fertility-sparing treatment of endometrial cancer. Am J Obstet Gynecol. 2021;224:370.e1–e13.
52. Vasan N, Cantley LC. At a crossroads: how to translate the roles of PI3K in oncogenic and metabolic signalling into improvements in cancer therapy. Nat Rev Clin Oncol. 2022;19:471–85.
53. Punekar SR, Velcheti V, Neel BG, Wong KK. The current state of the art and future trends in RAS-targeted cancer therapies. Nat Rev Clin Oncol. 2022;19:637–55.
54. Mullen J, Kato S, Sicklick JK, Kurzrock R. Targeting ARID1A mutations in cancer. Cancer Treat Rev. 2021;100:102287.
55. Gibson WJ, Hoivik EA, Halle MK, Taylor-Weiner A, Cherniack AD, Berg A, et al. The genomic landscape and evolution of endometrial carcinoma progression and abdominopelvic metastasis. Nat Genet. 2016;48:848–55.
56. Gotoh O, Sugiyama Y, Tonooka A, Kosugi M, Kitaura S, Minegishi R, et al. Genetic and epigenetic alterations in precursor lesions of endometrial endometrioid carcinoma. J Pathol. 2024;263:275–87.
57. Rim EY, Clevers H, Nusse R. The wnt pathway: from signaling mechanisms to synthetic modulators. Annu Rev Biochem. 2022;91:571–98.
58. Zhang L, Shay JW. Multiple roles of APC and its therapeutic implications in colorectal cancer. J Natl Cancer Inst. 2017;109:djw332.
59. Byron SA, Gartside M, Powell MA, Wellens CL, Gao F, Mutch DG, et al. FGFR2 point mutations in 466 endometrioid endometrial tumors: relationship with MSI, KRAS, PIK3CA, CTNNB1 mutations and clinicopathological features. PLoS One. 2012;7:e30801.
60. Giannakis M, Hodis E, Jasmine MX, Yamauchi M, Rosenbluh J, Cibulskis K, et al. RNF43 is frequently mutated in colorectal and endometrial cancers. Nat Genet. 2014;46:1264–6.
61. Joerger AC, Fersht AR. The p53 pathway: origins, inactivation in cancer, and emerging therapeutic approaches. Annu Rev Biochem. 2016;85:375–404.
62. Hernández Borrero LJ, El-Deiry WS. Tumor suppressor p53: biology, signaling pathways, and therapeutic targeting. Biochim Biophys Acta Rev Cancer. 2021;1876:188556.
63. Köbel M, Piskorz AM, Lee S, Lui S, LePage C, Marass F, et al. Optimized p53 immunohistochemistry is an accurate predictor of TP53 mutation in ovarian carcinoma. J Pathol Clin Res. 2016;2:247–58.
64. Vermij L, Léon-Castillo A, Singh N, Powell ME, Edmondson RJ, Genestie C, et al. p53 immunohistochemistry in endometrial cancer: clinical and molecular correlates in the PORTEC-3 trial. Mod Pathol. 2022;35:1475–83.
65. Vogelstein B, Kinzler KW. The path to cancer --three strikes and you're out. N Engl J Med. 2015;373:1895–8.

66. Tomasetti C, Marchionni L, Nowak MA, Parmigiani G, Vogelstein B. Only three driver gene mutations are required for the development of lung and colorectal cancers. Proc Natl Acad Sci USA. 2015;112:118–23.
67. Turajlic S, Swanton C. Metastasis as an evolutionary process. Science. 2016;352:169–75.
68. Birkbak NJ, McGranahan N. Cancer genome evolutionary trajectories in metastasis. Cancer Cell. 2020;37:8–19.
69. Moukarzel LA, Da Cruz PA, Ferrando L, Hoang T, Sebastiao APM, Pareja F, et al. Clonal relationship and directionality of progression of synchronous endometrial and ovarian carcinomas in patients with DNA mismatch repair-deficiency associated syndromes. Mod Pathol. 2021;34:994–1007.
70. Takahashi K, Yachida N, Tamura R, Adachi S, Kondo S, Abe T, et al. Clonal origin and genomic diversity in lynch syndrome-associated endometrial cancer with multiple synchronous tumors: identification of the pathogenicity of MLH1 p.L582H. Genes Chromosomes Cancer. 2024;63:e23231.
71. Moreno-Moreno E, Caniego-Casas T, Carretero-Barrio I, Cortés A, Muriel A, Domínguez-Rullán JA, et al. Histologic and molecular type changes in endometrial cancer recurrences in comparison with their corresponding primary tumors. Am J Surg Pathol. 2024;48:1580–7.

3 Hereditary Aspects of Endometrial Cancer

Takashi Mitamura

Abstract

Up to 5% of cancer patients fall under hereditary tumor syndromes, and they are in a state of high cancer susceptibility, mainly due to autosomal dominant inheritance. For some hereditary tumors, it has become possible to propose new preventive measures for pre-onset cancer (primary prevention to inhibit the onset of cancer, and secondary prevention to detect cancer early and prevent it from becoming severe) to people who have developed cancer or have not yet developed cancer, based on the guidelines. Women with hereditary cancer syndromes have an increased risk of developing endometrial cancer, which is mainly associated with germline heterozygous pathogenic variants of DNA mismatch repair genes of Lynch syndrome and PTEN of Cowden syndrome. It is predicted that the number of patients diagnosed with hereditary tumor syndromes will simultaneously increase because the number of treatments for which testing of mismatch repair protein deficiency or microsatellite instability will be used as companion diagnostics will increase in the future. Growing evidence indicates that diagnostic genetic testing is useful to find asymptomatic carriers of pathogenic variants and propose appropriate surveillance for secondary prevention. Conversely, there is still limited evidence for primary prevention of hereditary endometrial cancer.

Keywords

Hereditary cancer syndrome · Mismatch repair · Lynch syndrome · Cowden syndrome · Surveillance

T. Mitamura (✉)
Department of Obstetrics and Gynecology, Hokkaido University Graduate School of Medicine, Sapporo, Japan
e-mail: takami@huhp.hokudai.ac.jp

H. Watari, T. Mitamura (eds.), *Uterine Body Neoplasms*, Comprehensive Gynecology and Obstetrics, https://doi.org/10.1007/978-981-95-0946-1_3

3.1 Endometrial Cancer and Hereditary Cancer Syndromes

Women with hereditary cancer syndromes have an increased risk of developing endometrial cancer (EC), which is mainly associated with germline heterozygous pathogenic variants of DNA mismatch repair (MMR) genes (*MLH1*, *MSH2*, *MSH6*, *PMS2*) of Lynch syndrome and the tumor-suppressor gene phosphatase and tensin homolog (*PTEN*) of Cowden syndrome (or PTEN hamartoma tumor syndrome) acting as tumor suppressor genes (Table 3.1). The estimated average age at which EC is associated with these cancer predisposition syndromes is 5–10 years earlier than that of the general population. In addition, the cumulative risk throughout a lifetime is 6–27 times higher than that of the general population. Lynch syndrome pathogenic alleles are widely present in humans, with a 1:320 population frequency of a single allele [1]. Several previous studies have shown that the prevalence of endometrial cancer increases or decreases according to genes and variants [2], while recent evidence has shown that the lifetime risk of endometrial cancer by MMR genes is similar, except for *PMS2*, which has the lowest risk [3]. Carriers of pathogenic variants of MMR genes are at increased risk of MMR-deficient (MMRd) colon cancer, stomach cancer, urinary tract cancer, pancreatic cancer, bile duct cancer, small intestinal cancer, brain tumor, endometrial cancer, and ovarian cancer. Endometrial cancer is the most common type of cancer in women with Lynch syndrome. The prevalence of Cowden syndrome is estimated to be at least 1 in 200,000 [4]. Carriers of pathogenic variants of *PTEN* are at increased risk of breast cancer, thyroid cancer, renal cancer, melanoma, and endometrial cancer. Endometrial cancer is the second most common cancer among women. In general, a person with cancer predisposition syndromes, which are associated with the functional loss of tumor suppressor genes, has a pathogenic variant in one allele (autosomal dominant inheritance).

Table 3.1 Lynch syndrome: cancer risks

Gene	Estimated average age of presentation of EC	Cumulative risk for diagnosis through lifetime of EC	Estimated average age of presentation of OVCA	Cumulative risk for diagnosis through lifetime of OVCA
MLH1	49 years	34%–54%	46 years	4%–20%
MSH2 and EPCAM	47 – 48 years	21%–57%	43 years	8%–38%
MSH6	53–55 years	16%–49%	46 years	≤1%–13%
PMS2	49–50 years	13%–26%	51–59 years	1.3–3%
PTEN	45 years	28% - 50%	N/A	N/A
Japanese general population	55 – 59 years	2.1%	50 – 54 years	1.6%

According to Knudson's two-hit hypothesis, carcinogenic loss of function of associated tumor suppressor genes requires a somatic second hit that silences another functional allele. Regarding *MLH1*, somatic *MLH1* hypermethylation could be the second hit in *MLH1* silencing [5]. Another mechanism of Lynch syndrome is caused by heterozygous germline deletions of the 3′ end of *EPCAM*, located only 15 kb upstream of *MSH2*. These heritable deletions of *EPCAM* induce transcriptional read-through, a phenomenon that occurs when the transcription machinery fails to recognize the transcription termination site of *EPCAM* and continues transcribing beyond the gene boundaries into *MSH2*. Subsequent monoallelic hypermethylation of the *MSH2* promoter present on the same allele as the 3′ end *EPCAM* deletion results in *MSH2* silencing [6, 7]. Overall, endometrial cancer risk in *EPCAM* deletion carriers is lower than that in carriers of pathogenic variants in *MSH2*. The actual risk appears to depend on the size and location of the *EPCAM* deletion.

Constitutional mismatch repair deficiency (CMMRD) syndrome is a rare and aggressive cancer predisposition syndrome caused by biallelic (homozygous or compound heterozygous) germline pathogenic variants of the MMR genes [8]. The estimated penetrance of EC in CMMRD syndrome is relatively low, at less than 10%. However, a scoring system developed by the European Consortium Care for CMMRD to identify individuals who should undergo CMMRD genetic testing suggests that patients with EC under 25 years old should undergo germline testing [9].

In rare cases, germline biallelic pathogenic variants in *NTHL1* cause autosomal recessive tumor predisposition syndrome as NTHL1 tumor syndrome or NTHL1-related polyposis, and several studies have reported an increased risk of endometrial cancer [10, 11]. Several reports suggest that monoallelic *MUTYH* pathogenic variants may be associated with an increased risk of gastric, liver, breast, and endometrial cancers [12], whereas other reports have demonstrated no association with breast or endometrial cancers [3, 13, 14]. Multi-gene panel testing has suggested that these hereditary endometrial cancers account for up to 9% of all patients [15–17]. Although the spectrum of germline pathogenic variants occurring in endometrial cancer risk genes among people from certain racial and ethnic groups requires further research [3], a previous systematic review of patients with colorectal cancer suggested that multiethnic populations had the highest overall pooled prevalence (2.1%), followed by East Asia (1.8%), Europe (1.1%), and the Middle East (1.1%), suggesting that Lynch syndrome prevalence, which overwhelms other hereditary cancer syndromes causing endometrial cancer, across multiple ethnic, geographic, and clinical populations, is similar [18]. As such, patients with endometrial cancer who are identified as having an increased risk of these hereditary cancer syndromes should be offered genetic counseling [19].

3.2 Mechanisms

3.2.1 *MMR* [20] (Fig. 3.1)

MMR genes are involved in correcting errors that occur when DNA is copied in a cell, and the functional loss of MMR genes can lead to cancer. While the details of several steps in the complete MMR mechanism remain hypothetical, the molecular switch/sliding clamp model can explain why MSH2-MSH6 heterodimers recognize a mismatch and load multiple ATP-bound sliding clamps onto the DNA capable of bidirectional diffusion. This MMR initiation step appears to be common to all organisms. The mismatch search by MSH proteins is likely aided by the replicative processivity clamp β-clamp/PCNA. The component requirements of 3′- and 5′-directed excision reactions may differ, and the disposition of MLH1/PMS2 homologs that are associated with MSH sliding clamps shortly after being loaded onto the DNA by a mismatch appears to be different in the 5′ and 3′ excision reactions [20].

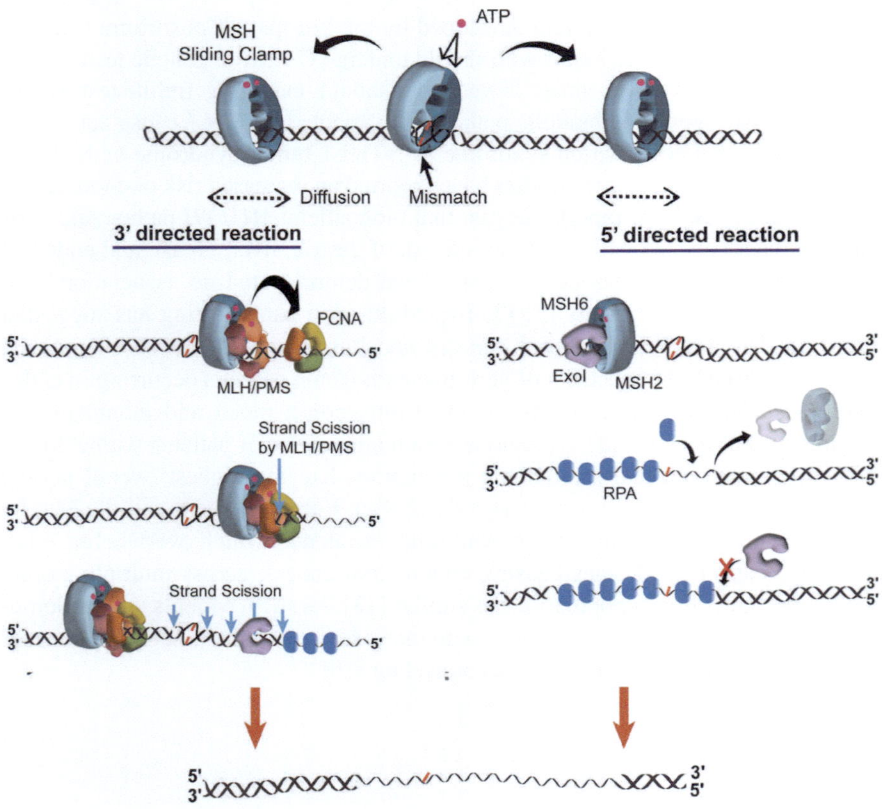

Fig. 3.1 The molecular switch/sliding clamp model for mismatch repair. (Reprinted with permission from Fishel and Lee [20])

The 5′ excision reaction is likely to occur on the lagging strand of a replication fork, where strand scission 5′ of a mismatch frequently occurs within each Okazaki fragment. It appears that this process relies on a physical interaction between the MSH sliding clamps and EXO1, which activates the 5′ → 3′ exonuclease activity of EXO1. Additionally, replication protein A (RPA) inhibits its resumption. In this model, the interaction and dissociation of MSH-EXO1 were proposed to be dynamic and redundant with excision termination. Normal S-phase replicative factors provide gap-filling and ligation. The 3′ excision reaction probably occurs on the leading strand of a replication fork and requires the MLH1/PMS2 endonuclease activity activated by PCNA. A hypothetical molecular switch/sliding clamp model for the 3′ excision imagines the MSH2/6-MLH1/PMS2 complex diffusing and interacting with PCNA associated with the 3′ end of the leading strand within the replication fork. This interaction activates the MLH1/PMS2 endonuclease to incise newly replicated DNA. Subsequently, EXO1 eliminates the error-containing strands. Since EXO1 is exclusively a 5′ → 3′ exonuclease, it is suggested that the 3′-directed MLH1/PMS2-PCNA activated endonuclease must present DNA fragments for EXO1 5′ → 3′ digestion. As with the previous excision reactions, this process seems dynamic and redundant, utilizing multiple MSH2/6-MLH1/PMS2 complexes until the mismatch is released and the loading of the MSH sliding clamps becomes refractory. Reassembling the S-phase replication machinery at the recessed 3′ end would restart DNA synthesis [20].

MMR deficiency refers to a genetic defect in MMR genes due to germline or somatic inactivation where the cell's ability to repair errors in DNA replication is impaired, resulting in an accumulation of mutations that can significantly increase the risk of developing cancer. This accumulation of mutations is characterized by the improperly repaired DNA sequences observed in microsatellites, the tandem repeats of 1–6 base pairs scattered throughout the genome in noncoding regions [21], and the so-called microsatellite instability (MSI). Overall, two clinical methods are primarily used to detect mismatch repair deficiency (MMRd) for companion diagnostic purposes: (1) immunohistochemistry (MMR-IHC) to assess tumor expression of MMR proteins; and (2) microsatellite instability (MSI) testing—typically PCR-based amplification of mononucleotide (e.g., BAT25, BAT26) or dinucleotide (e.g., D5S346, D2S123, D17S250) microsatellite markers that are polymorphic DNA loci containing repeated nucleotide sequences [22]. Tumor mutational burden (TMB)—the number of somatic mutations per megabase of tumor DNA—may be elevated in MMRd tumors and can serve as a supportive biomarker, but it is not a direct diagnostic test for MMR deficiency [23]. MSI is determined by comparing the length of nucleotide repeats in tumor cells and normal cells and can be analyzed by a fragment analysis using capillary electrophoresis. In general, the interpretation of MSI is classified into three categories according to the five NCI-recommended panels of microsatellite markers: MSI-high, defined as having instability in two or more markers; MSI-low, defined as having instability in one marker; and microsatellite stable (MSS), defined as lacking apparent instability in any of the markers [24]. Recent clinical data have demonstrated that the incidence of MSI-high in endometrial cancer is approximately 17% [25].

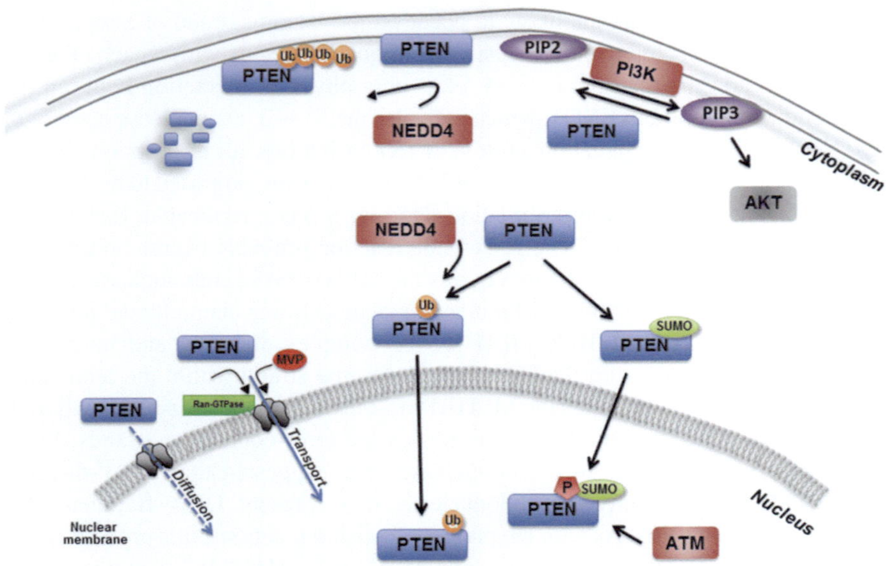

Fig. 3.2 Representative mechanisms of PTEN in the regulation of cell proliferation and cell growth. (Reprinted with permission from Papa and Pandolfi [26])

3.2.2 *PTEN* [26] (Fig. 3.2)

PTEN is a major tumor suppressor with both phosphatase-dependent and phosphatase-independent functions. As it dephosphorylates the second messenger phosphatidylinositol-3,4,5-triphosphate (PIP3), PTEN regulates the activation of the oncogenic PI3K-AKT signaling pathway and inhibits cell proliferation and cell growth. In addition, through protein phosphatase activity, PTEN can regulate the phosphorylation of protein substrates to inhibit cell migration and invasion. Recently, several studies have highlighted the importance of PTEN localization in cell nuclei. SUMOylation and NEDD4-dependent monoubiquitination favor PTEN shuttling and nuclear translocation. Carrier proteins, such as MVP or Ran-GTPase, transport PTEN into the nucleus. In addition, passive diffusion may further increase PTEN expression in the nucleus, where PTEN can function as a tumor suppressor in a phosphatase-independent manner [26].

Germline *PTEN* pathogenic variants can cause Cowden syndrome (or PTEN hamartoma tumor syndrome), a rare inherited disorder characterized by the presence of many benign growths called hamartomas and an increased risk of cancer. However, they lack robust evidence as biomarkers for specific molecular-targeted therapy and are not yet used for companion diagnostic tests, although they are sometimes found in cancer genome profiling tests and secondary findings.

3.2.3 MMRd Endometrial Cancer with Sporadic Causes

MLH1 promoter hypermethylation is the most common cause (64.3–72.4% [27, 28]) of MMRd endometrial cancer without germline pathogenic variants in an MMR gene, *MLH1*, *MSH2*, *MSH6*, and *PMS2*, or deletions in *EPCAM*, as characterized by loss of tumor MLH1 and PMS2 protein expression. Distinguishing between *MLH1* methylation and Lynch syndrome-related MMR deficiency is clinically critical for secondary cancer risk management and for identifying relatives at risk of cancer. Table 3.2 shows the correlation between endometrial cancer tissue testing by MMR-IHC, MSI, and *MLH1* promoter methylation and the suggested etiologies [3].

Lynch-like syndrome has similar clinicopathological features to Lynch syndrome-associated tumors due to loss of MMR protein expression without any identifiable pathogenic germline mismatch repair gene variant or MLH1 promoter hypermethylation. Lynch-like syndrome-associated tumors constitute up to 70% of cases with MSI-high and MMR deficiency suspected of involving Lynch syndrome [29]. The predominant cause of Lynch-like syndrome-related MMR deficiency, which is responsible for up to 80% of Lynch-like syndrome-associated cancers, involves double somatic mutations in the same MMR gene and outside of *MLH1* promoter methylation, known as biallelic MMR deficiency [29]. Another potential

Table 3.2 Tumor testing results and plausible etiologies

Immunohistochemistry				MSI	MLH1 promoter methylation	Plausible etiologies
MLH1	MSH2	MSH6	PMS2			
Positive	Positive	Positive	Positive	N/A	N/A	1. Sporadic cancer 2. Other hereditary cancer syndrome
Positive	Positive	Positive	Positive	MSI-H	N/A	1. Sporadic cancer 2. Germline PV in any of the LS genes
Negative	Positive	Positive	Negative	N/A	N/A	1. Sporadic cancer 2. Germline MLH1 PV or rarely PMS2
Negative	Positive	Positive	Negative	N/A	Positive	1. Sporadic cancer 2. Rarely germline MLH1 PV or constitutional MLH1 epimutation
Negative	Positive	Positive	Negative	N/A	Negative	1. Germline MLH1 PV or rarely PMS2 2. Sporadic cancer
Positive	Negative	Negative	Positive	N/A	N/A	1. Germline MSH2/EPCAM PV or rarely germline MSH6 PV 2. Sporadic cancer
Positive	Positive	Positive	Negative	N/A	N/A	1. Germline PMS2 PV 2. Germline MLH1 PV 3. Sporadic cancer
Positive	Negative	Positive	Positive	N/A	N/A	1. Germline MSH2/EPCAM PV 2. Sporadic cancer
Positive	Positive	Negative	Positive	N/A	N/A	1. Germline MSH6 PV 2. Germline MSH2 PV 3. Sporadic cancer
Negative	Positive	Positive	Positive	N/A	N/A	1. Sporadic cancer 2. Germline MLH1 PV 3. Germline PMS2 PV 4. Somatic MLH1 or PMS2 PV
Negative	Negative	Negative	Negative	N/A	N/A	1. Germline PV in any LS gene 2. Sporadic cancer

PV: pathogenic variants

mechanism that may underlie Lynch-like syndrome includes the presence of an atypical germline pathogenic variant in MMR genes not identified by current detection methods, or germline pathogenic variants in genes outside MMR genes [29]. In addition, germline mutations in genes other than those involved in the classical MMR system can mimic the Lynch syndrome phenotype with MMRd [30]. In clinical settings, a frequent explanation for Lynch-like syndrome cases that should always be ruled out is false-positive MMR-IHC/MSI results, which represent approximately 19% of cases in some series [31]. Therefore, the MMR-IHC and MSI statuses should be confirmed before classifying a case as a Lynch-like syndrome.

Several studies have shown that the risk of developing colorectal cancer and other Lynch syndrome-related cancers in Lynch-like syndrome cohorts and their first-degree relatives is lower than that in patients with Lynch syndrome. However, patients with Lynch-like syndrome are more likely to have related cancers than sporadic cases [30]. A previous study showed that the standardized incidence ratio of endometrial cancer in Lynch-like syndrome patients was 14.5, which was significantly lower than the value of 57.8 in Lynch syndrome patients [32]. In addition, the cumulative risk at 70 years was considerably lower (7.7%) in Lynch-like syndrome patients than in Lynch syndrome patients (36.1%, log-rank P=0.002 [32]) and higher than that in the general population (2.1% in Japan [33]).

3.3 The Diagnosis of Hereditary Endometrial Cancer

The ESGO guidelines describe that testing for the MMR-IHC/MSI in endometrial carcinoma patients is relevant for four reasons [1]: the diagnosis, as MMRd/MSI-high is considered a marker for endometrioid-type endometrial carcinoma [2]; pre-screening for identifying patients at higher risk for Lynch syndrome [3]; the prognosis, as identified by the Cancer Genome Atlas; and [4] determining the potential utility of immune checkpoint inhibitor therapy [19].

As shown in Table 3.2, MMRd and/or MSI-high can suggest germline pathogenic variants of MMR genes. However, in clinical practice, a previous study showed that more than 70% of MSI-high patients did not undergo confirmatory genetic testing when a next-generation sequencing (NGS)-based technique was used for the MSI analysis, in which MSI-high endometrial cancers were found in 15%, and only 24.44% of those who underwent the genetic test carried a pathogenic variant in a Lynch syndrome-associated gene [34]. An *MLH1* methylation analysis can eliminate more than 70% of the somatic origins of MMR deficiency [35]; however, an *MLH1* analysis is rarely requested in clinical practice.

3.4 Clinical Features of Hereditary Endometrial Cancer

The clinical and pathological features of Lynch syndrome-related endometrial cancer are critical for its treatment, surveillance, and prevention. However, clinicopathological information has been obtained from a limited number of patients because genetic testing rates among endometrial cancer patients are low (5–6% according to tumor registry data from 2013 to 2019 [3]). An older study suggested a high frequency of non-endometrioid or high-grade endometrioid types [36]. However, recent reports showed that the general characteristics of Lynch syndrome-related endometrial cancers were grade 1 or 2 (73.2–80.0%), endometrioid carcinoma (71.0–96.3%), and FIGO stage I or II (76.0–92.6%) [37–40]. In several studies [37, 38], the proportion of clear cell carcinoma was higher (11%–13.7%) than that of general endometrial carcinoma.

The prognosis for Lynch syndrome-related endometrial cancer is good. The Prospective Lynch Syndrome Database study showed 5- and 10-year crude survival rates of 136 patients with endometrial cancer and pathogenic *MLH1*, *MSH2*, and *MSH6* variants diagnosed after inclusion and before 65 years old, with 5- and 10-year survival rates of 89% and 89%, respectively [41]. According to the same Prospective Lynch Syndrome Database study, the mortality rate of patients with Lynch syndrome-related endometrial cancer and pathogenic *MLH1*, *MSH2*, and *MSH6* variants at 75 years old was 3% for all three genes [42]. Another Lynch syndrome database showed the 5- and 10-year overall survival rates for 104 patients, who were prospectively diagnosed at under 65 years of age, to be 93% and 93%, respectively [43]. In contrast, a retrospective study of three randomized controlled trials, the Post-Operative Radiation Therapy in Endometrial Carcinoma (PORTEC)-1, PORTEC-2, and PORTEC-3 trials, which have provided critical evidence for treating endometrial cancer, showed the prognostic impact of Lynch syndrome among MMRd endometrial cancers. The data from these PORTEC studies revealed that the estimated recurrence-free survival at 5 years was 83.7% for patients with Lynch syndrome-related MMRd endometrial cancer, 78.6% for patients with MLH1-methylated MMRd endometrial cancer, and 95.5% for patients with other causes of MMRd endometrial cancer [28]. These data demonstrate that the survival of Lynch syndrome-related endometrial cancer patients is comparable to the prognosis of general endometrial cancer patients. In addition, the former Prospective Lynch Syndrome Database study suggested that Lynch syndrome-related endometrial cancer was cured in most patients because the mid- and long-term survival rates were the same. Another study investigated the outcomes of patients with Lynch syndrome diagnosed with non-endometrioid endometrial cancer, and patients diagnosed with Lynch syndrome experienced better disease-free and overall survival than controls [44].

In endometrial cancer associated with Cowden syndrome, limited data suggest that endometrioid carcinoma is the predominant histological subtype [45]. Notably, sarcomas may occur in the uterus in a small number of patients (2.7%) [45]. As shown in Table 3.1, endometrial cancer occurs at an earlier age in Cowden syndrome than in other hereditary syndromes with a predisposition to endometrial cancer, and retrospective studies have shown that endometrial cancer diagnoses arise before 50 years old in up to 40% of Cowden syndrome patients [46].

3.5 Surveillance of Hereditary Endometrial Cancer

Surveillance of Lynch syndrome-related cancer has been suggested to effectively reduce the incidence and mortality of related cancers. For instance, a previous study showed that long-term colonoscopic surveillance did not significantly increase colon cancer mortality compared to that in pathogenic variant-negative relatives [47]. Regarding the surveillance for endometrial cancer, this study investigated the long-term effectiveness of endometrial suction biopsy and transvaginal ultrasonography that were offered beginning at 35 years old with surveillance visits at 2- to 3-year intervals in pathogenic variant-positive and pathogenic variant-negative Lynch syndrome family members. This surveillance of 55 at-risk women who did not complain of any symptoms and did not undergo prophylactic hysterectomy detected 17 women with endometrial cancer, and 13 patients (76.5%) had stage I disease [47]. Another prospective study showed that planned colonoscopic and gynecological surveillance resulted in few deaths, and the 10-year crude survival was 91% if the first cancer was colorectal, 98% if it was endometrial, and 89% if it was ovarian cancer [48]. However, more evidence is needed to determine the best age for the initiation of screening. Because early detection of endometrial cancer based on symptoms is also often possible, education regarding the importance of prompt reporting and evaluation of any abnormal uterine bleeding or postmenopausal bleeding is essential for carriers of hereditary cancer syndromes [3].

According to the latest NCCN guideline [3], in Lynch syndrome, screening via an endometrial biopsy, which is both highly sensitive and highly specific as a diagnostic procedure, can be considered every 1–2 years starting at 30–35 years old. The latest ESGO guidelines [19] propose that surveillance of the endometrium by annual transvaginal ultrasound and an annual or biennial biopsy should be considered in all Lynch syndrome mutation carriers from 35 years old. It is also recommended that clinicians consider tailored surveillance programs by considering individual factors to decide the starting age of surveillance and should integrate knowledge on the specific mutation and history of onset of events in the family.

In Lynch syndrome-related colon cancer, a previous study suggested that the prevalence of MMRd increased with the size of precancerous adenomatous polyps, and loss of the MMR function was a late event [49]. However, whether or not the size of the precancerous lesion, including atypical endometrial hyperplasia, correlates with MMRd endometrial cancer is unknown. Therefore, endometrial resection according to tumor size does not seem pragmatic. In addition, transvaginal ultrasound is not recommended as a screening tool in premenopausal patients because of the wide range of endometrial stripe thicknesses throughout the normal menstrual cycle [3]. In CMMRD syndrome, an annual pelvic examination with endometrial sampling is recommended, beginning as early as 20 years old [50]. In Cowden syndrome, screening via an endometrial biopsy every 1–2 years from 35 years old can be considered. In NTHL1 tumor syndrome, the same surveillance as in Lynch syndrome may be recommended for carriers of pathogenic variants, but the age to start is unknown.

3.6 Prophylaxis of Hereditary Endometrial Cancer

Chemoprevention

Regarding the prophylaxis of Lynch syndrome-related colon cancers, a previous randomized CAPP2 trial showed that participants taking daily aspirin for at least 2 years had a relative 35% reduction in the incidence of colorectal cancer compared to those who took a placebo (hazard ratio [HR], 0.65; 95% confidence interval [CI], 0.43–0.97; $P = 0.035$) after a mean 10-year follow-up [51]. Subsequently, some experts have recommended considering aspirin for chemoprevention [3]. Although there is no evidence of chemoprevention for Lynch syndrome-related endometrial cancer, retrospective studies have provided some suggestions. A systematic review suggested that the use of aspirin was associated with a reduced risk of endometrial cancer in cohort studies (pooled risk ratio = 0.86; 95% CI, 0.86–0.99), and the reduced risk was closely related to high frequency of use [52]. Although we do not know whether we can apply these data to Lynch syndrome carriers, some patients taking aspirin to prevent Lynch syndrome-related colorectal cancer might benefit from endometrial cancer reduction. Another retrospective cohort study suggested that the use of hormonal contraceptives was associated with a significantly lower risk of endometrial cancer compared with no contraceptive use in Lynch syndrome (HR, 0.39) [53].

Risk-Reducing Surgery

Because further randomized controlled trials are needed to confirm the benefits of both aspirin and contraceptives in the prevention of endometrial cancer, hysterectomy can be a risk-reducing option to reduce the risk of cancer and cancer mortality [54–56]. A previous representative study showed that among 61 women with germline mutations related to Lynch syndrome who had undergone prophylactic hysterectomy, there were no occurrences of endometrial cancer [57]. The NCCN guidelines recommend that the timing of hysterectomy be individualized based on whether childbearing is complete, comorbidities, and family history [3]. Because *PMS2* carriers appear to be at only a modestly increased risk of EC compared to *MLH1*, *MSH2*, and *MSH6* carriers, hysterectomy should also be individualized based on Lynch syndrome-related pathogenic variants [3].

In young patients with low-grade endometrioid cancer estimated to be in preoperative stage IA who strongly wish to preserve their ovaries, ovarian preservation after fully explaining the risks or recurrence is sometimes proposed because no significant differences in mortality or recurrence were found between the ovarian preservation group and the bilateral salpingo-oophorectomy group in previous studies [58]. In contrast, Lynch syndrome carriers are also at risk of ovarian cancer (Table 3.1), and bilateral salpingo-oophorectomy (BSO) may reduce the incidence of ovarian cancer. The decision to use BSO as a risk-reducing option in Lynch syndrome carriers without endometrial cancer or as a combined resection for therapeutic hysterectomy in endometrial cancer patients should be individualized. For example, among carriers bearing *MLH1* and *MSH6* pathogenic variants, hysterectomy with bilateral salpingectomy may be considered starting at 40 years old, with delayed bilateral oophorectomy starting at 50 years old. Given the increased risks of endometrial and ovarian cancers in carriers of *MSH2* pathogenic variants, hysterectomy with BSO may be considered as early as 40 years old. In carriers bearing *PMS2* pathogenic variants, hysterectomy with BSO should be considered after 50 years old. In addition, coordination between hysterectomy and oophorectomy should be discussed in patients requiring colorectal surgery [3]. In particular, in patients who undergo multiple surgeries to treat recurrences of colorectal cancer, serious pelvic adhesion may be an obstacle to subsequent hysterectomy. Furthermore, the ESGO guidelines recommend that hysterectomy and bilateral salpingo-oophorectomy to prevent Lynch syndrome-related endometrial and ovarian cancers be performed after childbearing and preferably before 40 years old [19]. The decision concerning prophylactic surgery needs to be discussed, including the risk of occult gynecological cancer detection. In addition, estrogen replacement therapy should be suggested if BSO is performed in premenopausal women.

A recent retrospective cohort study, in which risk-reducing surgery was performed in 73 carriers (median age, 51 years old), reported that 2 asymptomatic endometrial cancers were found, and all were diagnosed with FIGO stage I [59]. In terms of cost-effectiveness, a systematic review found that a strategy of surveillance alone or offering surveillance and risk-reducing surgery was cost-effective, except for Lynch syndrome with pathogenic variants of *PMS2*, and that offering only risk-reducing surgery was less effective than surveillance with or without surgery [56]. A previous study suggested that, for cancer prevention, 28 and 6 prophylactic surgeries would need to be performed to prevent 1 case of ovarian and endometrial cancers associated with Lynch syndrome, respectively [60]. However, a more recent study suggested that a very high cure rate after endometrial and ovarian cancers may call into question the place of risk-reducing hysterectomy with or without oophorectomy [48]. Individual risk estimation should be made by experts based on mutated genes, sex, and age, and the decision of general indication for risk-reducing surgery requires further study. Hysterectomy should also be considered for women with Cowden syndrome at the completion of childbearing [61], although no prospective evidence exists that compares this management strategy to surveillance alone.

Vaccines

Recent technological advances in tumor molecular profiling and a better understanding of genetic risk factors have transformed cancer prevention approaches, including cancer vaccines, for at-risk individuals. Mutations in coding microsatellites give rise to frameshift peptides (FSPs) that the immune system recognizes as neoantigens. Owing to clonal evolution, Lynch syndrome-related tumors share a set of recurrent and predictable FSP neoantigens in the same and different Lynch syndrome patients. Cancer vaccines composed of commonly recurring FSP neoantigens selected through prediction algorithms have been clinically evaluated in Lynch syndrome carriers and have proven to be safe and immunogenic. Preclinically analogous FSP vaccines have been shown to elicit FSP-directed immune responses and exert tumor-preventive effects in murine models of Lynch syndrome. Theoretically, the tumor spectrum in

Lynch syndrome encompasses organs such as the brain, skin, and kidney tumors in addition to Lynch syndrome-related primary tumors of the colorectum and endometrium, and the immunopreventive potential of neoantigens can be shared by tumors in different organs. Immunopreventive vaccines with these shared neoantigens that are further adjusted by an individual's HLA genotype may allow the development of personalized, preventive vaccines without organ restriction [62].

3.7 Secondary Findings of Cancer Genome Profiling Tests

Causative genetic changes (variants) are involved in the development of malignant tumors. Somatic variants usually turn on the switch that changes normal cells into tumor cells; however, in some cases, germline variants that the individual is born with may be present as a background. These germline variants may be associated with hereditary cancer. Therefore, obtaining a thorough family history at an early stage of medical treatment is important, and genetic counseling is proposed if hereditary diseases are predicted. Conversely, in the cancer comprehensive genomic profiling (CGP) test, which searches for treatment drugs for intractable malignant diseases according to the tumor variants, a large number of genes are comprehensively analyzed using next-generation sequencing. Therefore, the presence of hereditary diseases that are difficult to predict based on a patient's medical and family history may be suggested. In recent years, secondary findings have become generally known as useful in the health management of patients and their relatives, even though they are not the primary aim. As CGP testing is penetrating the daily medical treatment of malignant tumors, it is predicted that the number of opportunities to confirm a diagnosis of hereditary diseases based on secondary findings will increase in the future.

Two types of CGP tests are available: a tumor tissue-only test that uses only tumor tissue and a tumor/normal tissue-paired test that uses both tumor tissue and a blood sample. Both the tissue and blood samples contained nucleic acids from the patient's non-tumor cells. Therefore, all CGP tests could potentially detect pathogenic germline variants in each gene. In the tumor tissue-only test, the variants detected are a mixture of somatic and germline variants, so at that point, they are presumed to be germline pathogenic variants (PGPVs), and confirmation testing is required to determine whether or not the variant exists in the germline. For example, regarding *MLH1*, *MSH2*, *MSH6*, and *PMS2* of MMR genes, the authors disclosed PGPVs when patients had MSI-high tumors, and the variant allele frequency was more than 10%. In addition, *EPCAM* pathogenic variants, which indirectly affect *MSH2* function, and *PTEN* pathogenic variants are disclosed when patients have typical phenotypes and their variant allele frequency is more than 30% and 20% in single base pair substitutions and indels, respectively. In the author's experience with 2291 patients who underwent CGPs (OncoGuide™ NCC Oncopanel System [Sysmex Corporation], FoundationOne® CDx cancer genome profiling [Chugai Pharmaceutical], FoundationOne® Liquid CDx cancer genome profiling [Chugai Pharmaceutical], and GenMineTOP [Konica Minolta]) to identify promising drugs suggested by genetic alterations (the proportion of endometrial cancer was 2%, Fig. 3.3a), total germline pathogenic variants (GPVs) and PGPVs of MMR genes or *PTEN* were identified in up to 1.1% of patients. This total proportion is relatively low, but GPV and PGPV of *MSH2*, *MSH6*, and *PMS2* found in endometrial cancer by CGPs could be an important prompt (3.7%, 7.4%, and 3.7% of all GPV/PGPV cases, respectively, Fig. 3.3b) to check for hereditary endometrial cancer and perform suitable surveillance or prophylactic surgery.

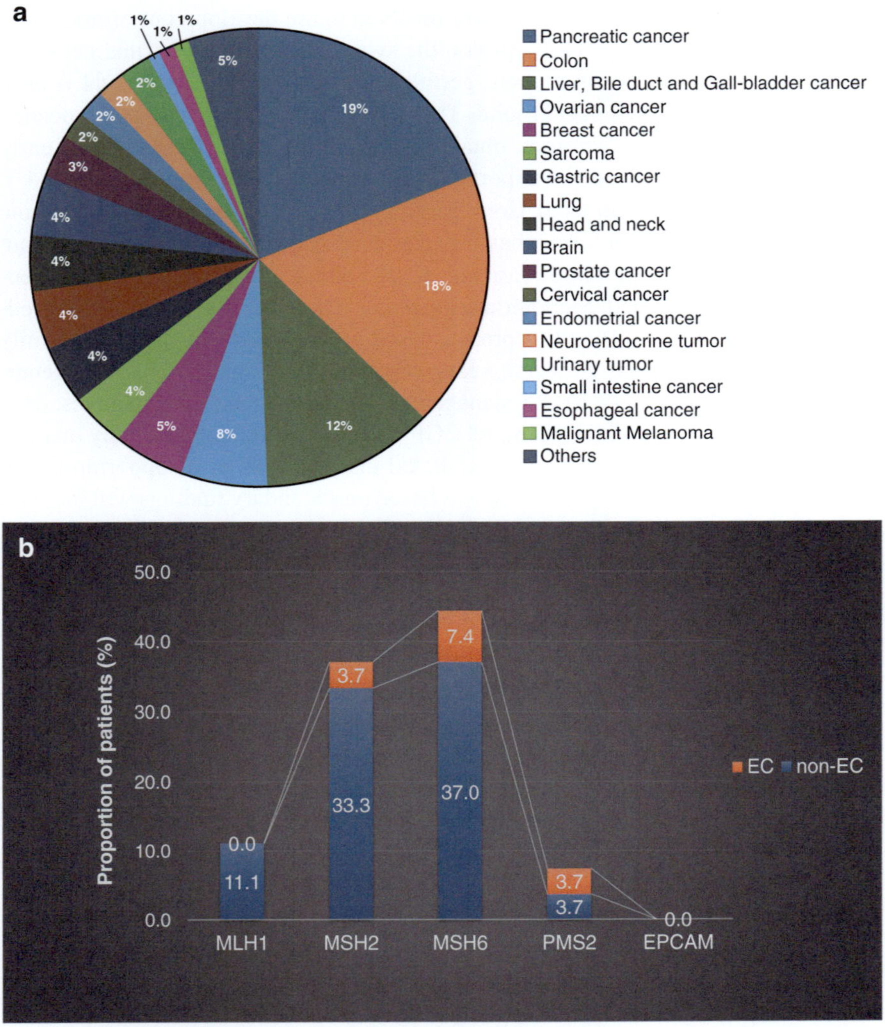

Fig. 3.3 Secondary findings in cancer comprehensive genomic profiling (CGP). (**a**) Tumor types and the proportions of CGP tests ($n = 2291$). (**b**) Proportion of germline pathogenic variants or presumed germline pathogenic variants of MSH2, MSH6, and PMS2 found in endometrial cancer

Whether or not to request a confirmatory test after obtaining secondary findings related to CGP tests is a matter of personal preference. Although the main genes to be disclosed have preventive measures and surveillance for their onset and are useful for health management, there are cases in which patients do not want to undergo genetic testing because of concerns about the psychological impact on relatives and genetic discrimination. Explaining that legal policies have been established to protect genetic information and prevent genetic discrimination and that progress is being made in establishing a social system that allows many citizens to reap the benefits of clinical genetics may help alleviate concerns. In addition, genetic counseling helps with the decision-making regarding whether or not to proceed with germline genetic testing, and it is strongly recommended for endometrial cancer patients and their families who are suspected of having hereditary tumor syndrome based on the secondary findings of CGP testing.

3.8 Future Directions

MMR-IHC/MSI testing can be used as a companion diagnostic test for the treatment of endometrial cancer with ICIs [63] (see Chap. 9), and the number of patients who undergo these tests will increase in the future. Simultaneously, the number of patients diagnosed with hereditary tumor syndromes will also increase. In such cases, clinicians need to maintain awareness of hereditary endometrial cancer so that patients who are unable to receive appropriate genetic medical care do not appear because only the treatment of the primary disease is prioritized. It is necessary to establish cooperation between malignant tumor treatment and genetic medical care in each region. In addition, tracking patients and their families with hereditary tumor syndromes through genetic medical care will increase opportunities to provide important treatment measures, including risk-reducing surgery and surveillance, at the appropriate time.

References

1. Roudko V, Cimen Bozkus C, Greenbaum B, Lucas A, Samstein R, Bhardwaj N. Lynch syndrome and MSI-H cancers: from mechanisms to "off-the-shelf" cancer vaccines. Front Immunol. 2021;12:757804.
2. Lynch HT, Riegert-Johnson DL, Snyder C, et al. Lynch syndrome-associated extracolonic tumors are rare in two extended families with the same EPCAM deletion. Am J Gastroenterol. 2011;106(10):1829–36.
3. National Comprehensive Cancer Network. Genetic/familial high-risk assessment: colorectal, endometrial, and gastric (version 3.2024). 2024. https://www.nccn.org/professionals/physician_gls/pdf/genetics_ceg.pdf. Accessed 4 Nov 2024.
4. Schultz KAP, Rednam SP, Kamihara J, et al. PTEN, DICER1, FH, and their associated tumor susceptibility syndromes: clinical features, genetics, and surveillance recommendations in childhood. Clin Cancer Res. 2017;23(12):e76–82.
5. Carnevali IW, Cini G, Libera L, et al. MLH1 promoter methylation could be the second hit in Lynch syndrome carcinogenesis. Genes (Basel). 2023;14(11):2060.

6. Ligtenberg MJ, Kuiper RP, Chan TL, et al. Heritable somatic methylation and inactivation of MSH2 in families with Lynch syndrome due to deletion of the 3′ exons of TACSTD1. Nat Genet. 2009;41(1):112–7.
7. Ligtenberg MJ, Kuiper RP, Geurts van Kessel A, Hoogerbrugge N. EPCAM deletion carriers constitute a unique subgroup of Lynch syndrome patients. Fam Cancer. 2013;12(2):169–74.
8. Ercan AB, Aronson M, Fernandez NR, et al. Clinical and biological landscape of constitutional mismatch-repair deficiency syndrome: an international replication repair deficiency consortium cohort study. Lancet Oncol. 2024;25(5):668–82.
9. Aronson M, Colas C, Shuen A, et al. Diagnostic criteria for constitutional mismatch repair deficiency (CMMRD): recommendations from the International Consensus Working Group. J Med Genet. 2022;59(4):318–27.
10. Beck SH, Jelsig AM, Yassin HM, Lindberg LJ, Wadt KAW, Karstensen JG. Intestinal and extraintestinal neoplasms in patients with NTHL1 tumor syndrome: a systematic review. Fam Cancer. 2022;21(4):453–62.
11. Weren RD, Ligtenberg MJ, Kets CM, et al. A germline homozygous mutation in the base-excision repair gene NTHL1 causes adenomatous polyposis and colorectal cancer. Nat Genet. 2015;47(6):668–71.
12. Win AK, Reece JC, Dowty JG, et al. Risk of extracolonic cancers for people with biallelic and monoallelic mutations in MUTYH. Int J Cancer. 2016;139(7):1557–63.
13. Thompson AB, Sutcliffe EG, Arvai K, et al. Monoallelic MUTYH pathogenic variants ascertained via multi-gene hereditary cancer panels are not associated with colorectal, endometrial, or breast cancer. Fam Cancer. 2022;21(4):415–22.
14. Fulk K, LaDuca H, Black MH, et al. Monoallelic MUTYH carrier status is not associated with increased breast cancer risk in a multigene panel cohort. Fam Cancer. 2019;18(2):197–201.
15. Cadoo KA, Mandelker DL, Mukherjee S, et al. Understanding inherited risk in unselected newly diagnosed patients with endometrial cancer. JCO Precis Oncol. 2019;3 https://doi.org/10.1200/po.18.00338.
16. Ring KL, Garcia C, Thomas MH, Modesitt SC. Current and future role of genetic screening in gynecologic malignancies. Am J Obstet Gynecol. 2017;217(5):512–21.
17. Karpel HC, Chern JY, Smith JM, Smith AJ, Pothuri B. Utility of germline multi-gene panel testing in patients with endometrial cancer. Gynecol Oncol. 2022;165(3):546–51.
18. Abu-Ghazaleh N, Kaushik V, Gorelik A, Jenkins M, Macrae F. Worldwide prevalence of Lynch syndrome in patients with colorectal cancer: systematic review and meta-analysis. Genet Med. 2022;24(5):971–85.
19. Concin N, Matias-Guiu X, Vergote I, et al. ESGO/ESTRO/ESP guidelines for the management of patients with endometrial carcinoma. Int J Gynecol Cancer. 2021;31(1):12–39.
20. Fishel R, Lee J-B. Mismatch repair. In: Hanaoka F, Sugasawa K, editors. DNA replication, recombination, and repair: molecular mechanisms and pathology. Tokyo: Springer Japan; 2016. p. 305–39.
21. Hamada H, Petrino MG, Kakunaga T. A novel repeated element with Z-DNA-forming potential is widely found in evolutionarily diverse eukaryotic genomes. Proc Natl Acad Sci USA. 1982;79(21):6465–9.
22. Boland CR, Thibodeau SN, Hamilton SR, et al. A national cancer institute workshop on microsatellite instability for cancer detection and familial predisposition: development of international criteria for the determination of microsatellite instability in colorectal cancer. Cancer Res. 1998;58(22):5248–57.
23. Snyder A, Makarov V, Merghoub T, et al. Genetic basis for clinical response to CTLA-4 blockade in melanoma. N Engl J Med. 2014;371(23):2189–99.
24. Umar A, Boland CR, Terdiman JP, et al. Revised Bethesda guidelines for hereditary nonpolyposis colorectal cancer (Lynch syndrome) and microsatellite instability. J Natl Cancer Inst. 2004;96(4):261–8.
25. Akagi K, Oki E, Taniguchi H, et al. Real-world data on microsatellite instability status in various unresectable or metastatic solid tumors. Cancer Sci. 2021;112(3):1105–13.

26. Papa A, Pandolfi PP. Phosphatase-independent functions of the tumor suppressor PTEN. In: Neel BG, Tonks NK, editors. Protein tyrosine phosphatases in cancer. New York: Springer New York; 2016. p. 247–60.
27. Buchanan DD, Tan YY, Walsh MD, et al. Tumor mismatch repair immunohistochemistry and DNA MLH1 methylation testing of patients with endometrial cancer diagnosed at age younger than 60 years optimizes triage for population-level germline mismatch repair gene mutation testing. J Clin Oncol. 2014;32(2):90–100.
28. Post CCB, Stelloo E, Smit V, et al. Prevalence and prognosis of Lynch syndrome and sporadic mismatch repair deficiency in endometrial cancer. J Natl Cancer Inst. 2021;113(9):1212–20.
29. Nugroho PP, Ghozali SAS, Buchanan DD, Pisano MI, Reece JC. Risk of cancer in individuals with Lynch-like syndrome and their families: a systematic review. J Cancer Res Clin Oncol. 2023;149(1):25–46.
30. Martinez-Roca A, Giner-Calabuig M, Murcia O, et al. Lynch-like syndrome: potential mechanisms and management. Cancers (Basel). 2022;14(5):1115.
31. Haraldsdottir S, Hampel H, Tomsic J, et al. Colon and endometrial cancers with mismatch repair deficiency can arise from somatic, rather than germline, mutations. Gastroenterology. 2014;147(6):1308–16 e1.
32. Bucksch K, Zachariae S, Aretz S, et al. Cancer risks in Lynch syndrome, Lynch-like syndrome, and familial colorectal cancer type X: a prospective cohort study. BMC Cancer. 2020;20(1):460.
33. Cancer statistics in Japan. https://ganjoho.jp/reg_stat/statistics/stat/summary.html.
34. Papadopoulou E, Rigas G, Fountzilas E, et al. Microsatellite instability is insufficiently used as a biomarker for Lynch syndrome testing in clinical practice. JCO Precis Oncol. 2024;8:e2300332.
35. Hampel H, Pearlman R, de la Chapelle A, et al. Double somatic mismatch repair gene pathogenic variants as common as Lynch syndrome among endometrial cancer patients. Gynecol Oncol. 2021;160(1):161–8.
36. Carcangiu ML, Radice P, Casalini P, Bertario L, Merola M, Sala P. Lynch syndrome—related endometrial carcinomas show a high frequency of nonendometrioid types and of high FIGO grade endometrioid types. Int J Surg Pathol. 2010;18(1):21–6.
37. Bounous VE, Robba E, Perotto S, et al. Gynecological cancers in Lynch syndrome: a comparison of the histological features with sporadic cases of the general population. J Clin Med. 2022;11(13):3689.
38. Yang Z, Liu X, Yang X, Liao QP. Screening and identification of Lynch syndrome: a systematic review of the frequency of Lynch syndrome-associated clinicopathologic and molecular characteristics in Lynch syndrome gynecologic cancers. Transl Cancer Res. 2021;10(10):4523–31.
39. Bartosch C, Pires-Luís AS, Meireles C, et al. Pathologic findings in prophylactic and nonprophylactic hysterectomy specimens of patients with Lynch syndrome. Am J Surg Pathol. 2016;40(9):1177–91.
40. Ramchander NC, Ryan NAJ, Walker TDJ, et al. Distinct immunological landscapes characterize inherited and sporadic mismatch repair deficient endometrial cancer. Front Immunol. 2019;10:3023.
41. Dominguez-Valentin M, Sampson JR, Seppala TT, et al. Cancer risks by gene, age, and gender in 6350 carriers of pathogenic mismatch repair variants: findings from the prospective Lynch syndrome database. Genet Med. 2020;22(1):15–25.
42. Dominguez-Valentin M, Haupt S, Seppala TT, et al. Mortality by age, gene and gender in carriers of pathogenic mismatch repair gene variants receiving surveillance for early cancer diagnosis and treatment: a report from the prospective Lynch syndrome database. EClinicalMedicine. 2023;58:101909.
43. Moller P, Seppala TT, Bernstein I, et al. Cancer risk and survival in path_MMR carriers by gene and gender up to 75 years of age: a report from the prospective Lynch syndrome database. Gut. 2018;67(7):1306–16.
44. Bogani G, Tibiletti MG, Ricci MT, et al. Lynch syndrome-related non-endometrioid endometrial cancer: analysis of outcomes. Int J Gynecol Cancer. 2020;30(1):56–61.

45. Mahdi H, Mester JL, Nizialek EA, Ngeow J, Michener C, Eng C. Germline PTEN, SDHB-D, and KLLN alterations in endometrial cancer patients with Cowden and Cowden-like syndromes: an international, multicenter, prospective study. Cancer. 2015;121(5):688–96.
46. Manning-Geist BL, Gatius S, Liu Y, et al. Diagnosis and management of an endometrial cancer patient with Cowden syndrome. Gynecol Oncol. 2021;163(1):14–21.
47. Jarvinen HJ, Renkonen-Sinisalo L, Aktan-Collan K, Peltomaki P, Aaltonen LA, Mecklin JP. Ten years after mutation testing for Lynch syndrome: cancer incidence and outcome in mutation-positive and mutation-negative family members. J Clin Oncol. 2009;27(28):4793–7.
48. Moller P, Seppala T, Bernstein I, et al. Cancer incidence and survival in Lynch syndrome patients receiving colonoscopic and gynaecological surveillance: first report from the prospective Lynch syndrome database. Gut. 2017;66(3):464–72.
49. Yurgelun MB, Goel A, Hornick JL, et al. Microsatellite instability and DNA mismatch repair protein deficiency in Lynch syndrome colorectal polyps. Cancer Prev Res (Phila). 2012;5(4):574–82.
50. Durno C, Boland CR, Cohen S, et al. Recommendations on surveillance and management of biallelic mismatch repair deficiency (BMMRD) syndrome: a consensus statement by the US multi-society task force on colorectal cancer. Gastroenterology. 2017;152(6):1605–14.
51. Burn J, Sheth H, Elliott F, et al. Cancer prevention with aspirin in hereditary colorectal cancer (Lynch syndrome), 10-year follow-up and registry-based 20-year data in the CAPP2 study: a double-blind, randomised, placebo-controlled trial. Lancet. 2020;395(10240):1855–63.
52. Wang Y, Zhao J, Chen X, Zhang F, Li X. Aspirin use and endometrial cancer risk: a meta-analysis and systematic review. Ann Transl Med. 2020;8(7):461.
53. Dashti SG, Chau R, Ouakrim DA, et al. Female hormonal factors and the risk of endometrial cancer in Lynch syndrome. JAMA. 2015;314(1):61–71.
54. Ryan N, Nobes M, Sedgewick D, Teoh SN, Evans DG, Crosbie EJ. A mismatch in care: results of a United Kingdom-wide patient and clinician survey of gynaecological services for women with Lynch syndrome. BJOG. 2021;128(4):728–36.
55. Duenas N, Navarro M, Teule A, et al. Assessing effectiveness of colonic and gynecological risk reducing surgery in Lynch syndrome individuals. Cancers (Basel). 2020;12(11):3419.
56. Snowsill TM, Coelho H, Morrish NG, et al. Gynaecological cancer surveillance for women with Lynch syndrome: systematic review and cost-effectiveness evaluation. Health Technol Assess. 2024;28(41):1–228.
57. Schmeler KM, Lynch HT, Chen LM, et al. Prophylactic surgery to reduce the risk of gynecologic cancers in the Lynch syndrome. N Engl J Med. 2006;354(3):261–9.
58. Morisada T, Kobayashi Y, Nagase S, et al. Japan Society of Gynecologic Oncology 2023 guidelines for treatment of uterine body neoplasm. J Gynecol Oncol. 2025;36(3):e119.
59. Eikenboom EL, van Doorn HC, Dinjens WNM, et al. Gynecological surveillance and surgery outcomes in Dutch Lynch syndrome carriers. Cancers (Basel). 2021;13(3):459.
60. Chen LM, Yang KY, Little SE, Cheung MK, Caughey AB. Gynecologic cancer prevention in Lynch syndrome/hereditary nonpolyposis colorectal cancer families. Obstet Gynecol. 2007;110(1):18–25.
61. National Comprehensive Cancer Network. Genetic/familial high-risk assessment: breast, ovarian, and pancreatic(version 2.2025). 2024. https://www.nccn.org/professionals/physician_gls/pdf/genetics_bopp.pdf. Accessed 6 Dec 2024.
62. Sei S, Ahadova A, Keskin DB, et al. Lynch syndrome cancer vaccines: a roadmap for the development of precision immunoprevention strategies. Front Oncol. 2023;13:1147590.
63. Westin SN, Moore K, Chon HS, et al. Durvalumab plus carboplatin/paclitaxel followed by maintenance durvalumab with or without olaparib as first-line treatment for advanced endometrial cancer: the Phase III DUO-E Trial. J Clin Oncol. 2024;42(3):283–99.

FIGO2023 Staging System and Molecular Classification for Endometrial Cancer

4

Hiroyuki Yamazaki and Hiroshi Asano

Abstract

Endometrial cancer is the most common gynecologic malignancy. To optimize treatment, avoiding overtreatment in low-risk patients and developing new therapies for high-risk cases are essential. The Cancer Genome Atlas (TCGA) identified four molecular clusters with distinct genomic features. The association between prognosis and the classification of molecular subtypes has been validated in multiple cohorts of endometrial cancer with various testing methods for surrogate markers. It is now incorporated into the WHO fifth edition classification using surrogate markers: *POLE*-mutated (*POLE*mut), mismatch repair-deficient (MMRd), no specific molecular profile (NSMP), and p53 abnormal (p53abn). Based on these findings, the 2023 FIGO staging system includes molecular subtypes for the first time. It also integrates pathological risk factors such as histological type, grade, lymphovascular space invasion (LVSI), and micrometastasis, highlighting the need for detailed evaluation. The relationship between molecular subtypes and prognosis is becoming central to treatment decisions. Molecular-based adjuvant therapies and novel treatments for advanced or recurrent disease are under development, with growing expectations for personalized care in endometrial cancer.

Keywords

Molecular classification · POLE · MSI/MMRd · p53abn · NSMP

H. Yamazaki (✉) · H. Asano
Hokkaido University, Sapporo, Japan

H. Watari, T. Mitamura (eds.), *Uterine Body Neoplasms*, Comprehensive Gynecology and Obstetrics, https://doi.org/10.1007/978-981-95-0946-1_4

4.1 Introduction

Endometrial cancer is the most common malignant tumor in gynecological cancer, and 80% of patients are in early-stage cancer and have a good prognosis. On the other hand, advanced and recurrent endometrial cancer has a poor prognosis. Thus, it has been a challenge to avoid excessive treatment for patients with good prognoses while at the same time developing novel treatment strategies for patients with poor prognoses.

Endometrial cancer has traditionally been classified based on morphological features, which has provided a fundamental understanding of tumor histology and progression. The conventional classification of endometrial cancer is based on estrogen dependency and well reflects morphological features: Type 1 is estrogen-dependent and well-differentiated. Type 2 is estrogen-independent and poorly differentiated [1]. However, it has been limited in predicting prognosis and guiding treatment plans for individual patients. The association between the prognosis and the classification of molecular subtypes for endometrial cancer, which originated from the comprehensive molecular and genomic analysis by the Cancer Genome Atlas (TCGA) [2], has been validated in multiple cohorts of endometrial cancer with various testing methods for surrogate markers. Furthermore, it has become clear that some pathological risk factors affect prognosis and/or lymph node metastasis, such as number of lymphovascular space invasion (LVSI) and micrometastasis to pelvic or para-aortic lymph node. Based on these findings, the International Federation of Gynecology and Obstetrics (FIGO) announced a revised uterine cancer staging system in 2023 [3]. In this revision, testing for the molecular prognostic factors is required in clinical practice and used for the first time in the staging system for endometrial cancer. In addition, pathological recurrent risk factors have been included in the staging system, and a more detailed evaluation is needed.

In this chapter, we will outline the clinical significance of the classification of molecular subtypes and describe the clinical features of each molecular subtype. We will also explain additional molecular genomic prognostic factors and then summarize the update points of the FIGO 2023 staging system, including pathological risk factors. We hope to help guide this chapter in the treatment plan for individual patients with endometrial cancer.

4.2 Overview of Molecular Classification for Endometrial Cancer

4.2.1 Comprehensive Molecular Analysis of TCGA

TCGA was launched in 2006 as a collaborative project between the National Cancer Institute and the National Human Genome Research Institute, aimed at collecting comprehensive genomic, epigenomic, transcriptomic, and proteomic data across various cancer types [4]. For endometrial cancer, TCGA planned an integrated genomic analysis based on comprehensive molecular genomics and reported the results of an integrated analysis in 2013 [2].

The TCGA analysis revealed that endometrial cancers could be divided into four integrated clusters based on multiplatform analysis: *POLE* (ultramutated), MSI (hypermutated), copy-number low (endometrioid), and copy-number high (serous-like).*POLE* cluster is mostly involved in cellular metabolism, which had the highest tumor mutational burden (TMB) of 232×10^{-6} mutations per Mb, increased C to A transversion frequency, and DNA polymerase epsilon (*POLE*) exonuclease domain mutations (EDM).MSI cluster was characterized by decreased *MLH1* mRNA expression, probably due to *MLH1* promoter methylation, and hypermutated tumors (18×10^{-6} mutations per Mb), resulting in microsatellite instability (MSI). Copy-number low (CN low) cluster had increased levels of progesterone receptor expression but low TMB (2.9×10^{-6} mutations per Mb) and somatic copy number alterations (SCNAs), consisting of most microsatellite stable (MSS) endometrioid cancers. Copy-number high (CN high) cluster was characterized by extensive SCNAs with a low TMB (2.3×10^{-6} mutations per Mb), and 90% of the cluster had *TP53* mutations.

These integrated genomic features were revolutionary as it stratified prognosis independent of conventional clinicopathological factors such as histological subtype, stage, and treatment strategy. That is, the prognosis of the *POLE* cluster was extremely good, that of the MSI and CN low clusters were intermediate, and that of the CN high cluster was extremely poor. Therefore, a classification method using surrogate markers that can be implemented in clinical practice was investigated.

4.2.2 Development of Surrogates for TCGA Molecular Cluster in Clinical Practice

In *Trans*PORTEC, an international consortium related to the PORTEC3 trial [5], *POLE* (ultramutated) in TCGA was replaced by a test using Sanger sequencing for hot spot mutations in exons 9 and 13 in *POLE*, MSI was replaced by a method using PCR, and CN high was replaced by immunohistochemistry (IHC) for the p53 protein and Sanger sequencing for *TP53* hot spot mutations. On the other hand, in the Proactive Molecular Risk Classifier for Endometrial Cancer (ProMisE) [6], Sanger sequencing for hot spot mutations in exons 9, 13, and 14 in *POLE* was used, but MSI and CN high in TCGA were replaced by IHC for mismatch repair (MMR)-related proteins and p53, respectively, with the criteria for p53-IHC testing reviewed. Although both studies differed in classification algorithms, prognostic stratification using these surrogate markers was verified in multiple cohorts.

4.2.3 WHO Molecular Subtype Classification

With the developments for clinical implementation described above, the fifth edition of the World Health Organization (WHO) Classification of Tumours published

Table 4.1 The testing methods for surrogate markers in WHO molecular classification

TCGA cluster	WHO molecular classification	Testing methods	Surrogate marker
POLE (ultramutated)	*POLE* mutation (*POLE*mut)	Sanger sequencing, NGS	*POLE*-EDM, including p.P286R, p.V411L, p.A456P, p.S297F, p.S459F
MSI (hypermutated)	MMR-deficient (MMR-d)	MSI testing by PCR or NGS	Microsatellite regions, including BAT25, BAT26, NR21, NR24, MONO27
		MMR-IHC testing by IHC	Mismatch repair (MMR)-related proteins, including MLH1, PMS2, MSH2, MSH6
CN high (serous-like)	p53 abnormal (p53abn)	Sanger sequencing, NGS	*TP53* pathogenic mutations
		IHC	p53 protein

EDM exonuclease domain mutations, *MSI* microsatellite instability, *MMR* mismatch repair, *POLE* DNA polymerase epsilon, *WHO* World Health Organization

in 2020 adopted a molecular subtype classification for endometrioid carcinoma of the uterine corpus [7]. The WHO classification mentions testing methods for surrogate markers in clinical practice (Table 4.1), and the algorithm for classification has been unified (Fig. 4.1). In this classification, endometrioid carcinoma is classified into the *POLE* pathogenic mutation (*POLE*mut), which corresponds to the *POLE* (ultramutated) in the TCGA; the MMR-deficient (MMRd), which corresponds to the MSI (hypermutated); the no specific molecular profile (NSMP), which corresponds to the CN low (endometrioid); and the p53 abnormal (p53abn), which corresponds to the CN high (serous-like). We will describe in detail the characteristics of each molecular subtype, clinically available testing methods, and surrogate markers in "3. Clinical Features of Each Molecular Subtype."

4.2.4 Clinical Significance of WHO Molecular Subtype Classification

Studies such as TCGA have examined the clinical prognosis of molecular subtypes, focusing on endometrioid carcinoma. In addition, most of the cohorts were treated with radiation therapy as adjuvant therapy. Therefore, the clinical significance of molecular classification in other histological subtypes and in patients where chemotherapy was selected as adjuvant therapy was also examined.

Hammer et al. investigated the relationship between molecular subtype classification and prognosis in high-grade endometrial cancer, especially uterine undifferentiated carcinoma and carcinosarcoma [8]. Even in those histological subtypes, the WHO molecular subtype classification was more critical in determining prognosis than the pathological recurrence risk classification. In addition, Kim et al. performed molecular subtype classification of clear cell carcinoma of the uterine corpus [9].

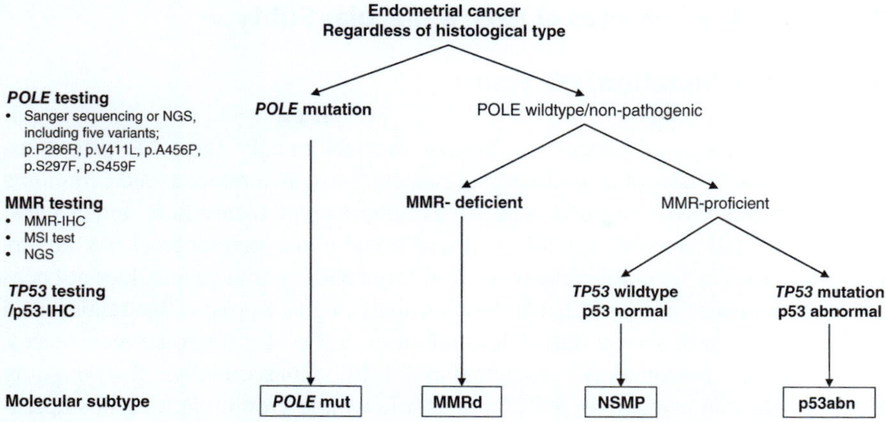

Fig. 4.1 Algorithm for molecular classification of endometrial cancer. Molecular classification of uterine cancer is appropriately performed by evaluating the results of each test in the order of *POLE* status, MMR status, and p53 status. FIGO2023 recommends that molecular classification be performed on all cases of endometrial cancer, regardless of histological type or stage. IHC immunohistochemistry, MMR mismatch repair, MMRd mismatch repair deficient, MSI microsatellite instability, NGS next-generation sequencing, NSMP no specific molecular profile, p53abn p53 abnormal, *POLE* DNA polymerase epsilon, *POLE*mut *POLE* mutation

They found that *POLE* and MMRd had a good prognosis, while p53abn had an extremely poor prognosis in clear cell carcinoma. However, NSMP had a high frequency of L1 cell adhesion molecule (L1CAM) overexpression and a poor prognosis similar to that of p53abn.

Yamazaki et al. investigated the relationship between molecular classification and prognosis when adjuvant chemotherapy was administered to patients with an intermediate or high risk of recurrence [10]. They reported that prognosis stratification is complex when the MMRd is extracted first according to the surrogate marker and algorithm of the ProMisE. Still, prognosis stratification is possible by extracting the *POLE*-EDM first, as in the WHO molecular subtype classification. This is related to the fact that the prognosis of *POLE*-EDM endometrial cancer patients who are treated with adjuvant chemotherapy is also excellent and that some cases with MMRd characteristics coexist with *POLE*-EDM. In addition, Asami et al. reported that prognosis stratification is possible by classifying molecular subtypes using a diagnostic algorithm according to the WHO molecular subtype classification with ProMisE surrogate markers [11]. These results indicate that the WHO molecular classification is essential for predicting prognosis, regardless of histological subtypes or differences in adjuvant therapy.

Based on the above, it is recommended that the FIGO 2023 staging system include the identification of the molecular subtype for all histological subtypes, not just endometrioid carcinoma.

4.3 Clinical Features of Each Molecular Subtype

4.3.1 *POLE* Mutation (*POLE*mut)

*POLE*mut endometrial cancer exhibits an overwhelmingly favorable prognosis, accounting for 8–10% of all endometrial cancers. Most are endometrioid carcinoma at early stages (I/II), respond well to treatment upon recurrence, and achieve extended overall survival, regardless of traditional clinicopathological risk factors [12–14]. Given its favorable prognosis, *POLE*mut endometrial cancer does not benefit from adjuvant therapy [12, 14]. Studies analyzing two prospective randomized trial databases have shown that *POLE*mut cases had no locoregional recurrences, even without postoperative radiotherapy [15]. Consequently, the ongoing PORTEC-4a trial categorizes *POLE*mut as a favorable group, omitting postoperative radiotherapy [16]. Although limited, reports indicate that advanced (stage III/IV) and recurrent cases also demonstrate good prognosis [12, 14].

The important genomic feature of *POLE*mut endometrial cancer is ultramutated, and it is essential to detect hot spot mutations in the exonuclease domain that affects the proofreading function of *POLE* [17]. To date, 11 variants in exons 9, 11, 13, and 14 have been investigated for their association with clinical prognosis: p.P286R, p.V411L, p.A456P, p.S297F, p.S459F, p.F367S, p.L424I, p.P436R, p.D368Y, p.M295R, and p.M444K [18, 19]. Leon-Castillo et al. used TCGA data to develop a scoring system for the pathogenicity of *POLE* in endometrial cancer based on genomic characteristics as variants to be detected in endometrial cancer. They reported that 11 variants described above were high POLE-score variants. Furthermore, McAlpine et al. investigated the prognosis of *POLE*mut endometrial cancer in PubMed from 2012 to 2018. When these 11 variants were defined as *POLE*mut, 294 of 399 cases with pathogenic mutations in *POLE* were diagnosed with *POLE*mut, and the prognosis was significantly better than other variants in *POLE*.

Figure 4.2 shows the frequency of pathogenic *POLE* variants in endometrial cancer registered in cBioPortal [20]. Pathogenic *POLE* variants were detected in 184 out of 5063 endometrial cancer samples, excluding duplications. The most common variant was p.P286R, accounting for 40% of pathogenic *POLE* variants in endometrial cancer. The sum of five variants defined by the fifth edition of WHO classification accounted for 85.9%, and that of 11 variants, including six with high POLE-score, accounted for 92.4%. There is a lack of data on clinical prognosis for other variants, and further consideration is needed to determine whether they should be included in the *POLE*mut.

Sanger sequencing, or next-generation sequencing (NGS), is the gold standard for detecting these variants. Alternative methods, such as quantitative polymerase chain reactions (PCR), have been reported [16]. Since each method's detection limit differs, it is necessary to select an appropriate method based on the tumor content and volume for individual patients.

Some of *POLE*mut endometrial cancer overlaps with other markers, shown in Fig. 4.3. In a study of cases undergoing extended surgery with lymphadenectomy,

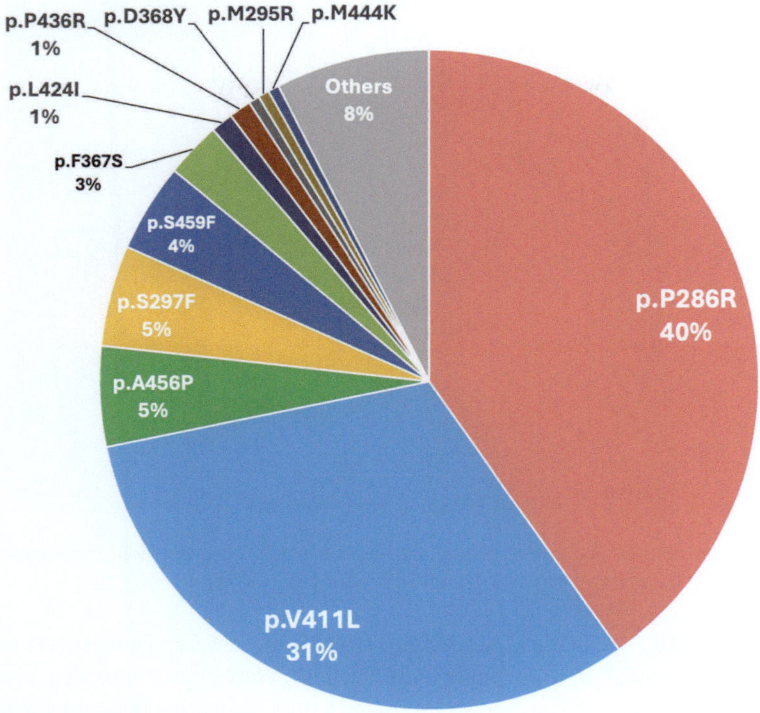

Fig. 4.2 Relative frequencies of pathogenic *POLE* variants in endometrial cancer. The pie chart shows the relative frequencies of pathogenic *POLE* variants detected in 184 out of 5063 endometrial cancer samples, excluding duplications in cBioPortal. (https://www.cbioportal.org/, Data acquisition date: January 3, 2025)

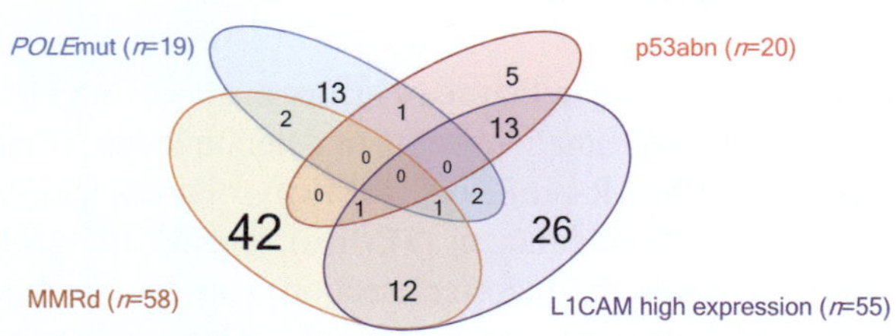

Fig. 4.3 Overlap of diagnostic markers in molecular classification of endometrial cancer. L1CAM L1 cell adhesion molecule, MMRd mismatch repair deficient, p53abn p53 abnormal, *POLE*mut DNA polymerase epsilon mutation

31.6% (6/19) overlapped with *POLE*mut, all showing favorable prognosis [10]. Therefore, extracting *POLE*mut first in molecular classification is deemed essential.

*POLE*mut tumors often present with high-grade endometrial cancer, such as grade 3 endometrioid carcinoma, and are characterized by high tumor-infiltrating lymphocytes (TIL), reflecting an ultramutated phenotype [21]. Fig. 4.4 shows a slide of hematoxylin and eosin (H&E) staining of a tumor with a pathogenic

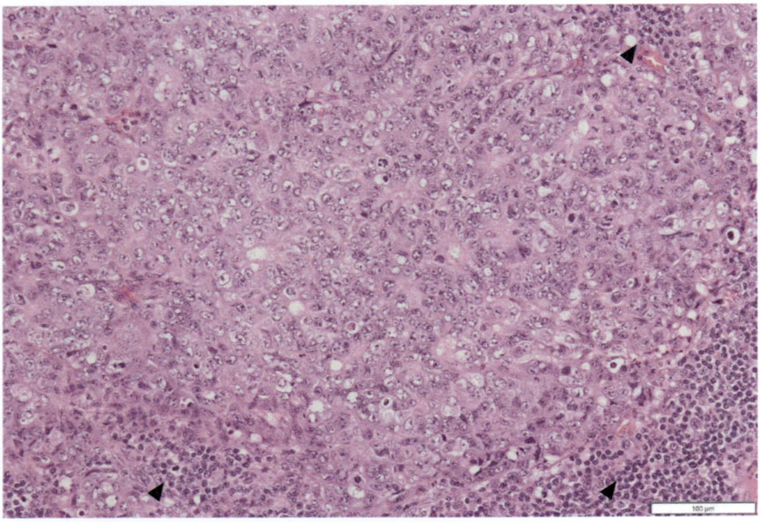

Fig. 4.4 Hematoxylin and eosin staining section of a grade-3 endometrioid carcinoma with a pathogenic *POLE* variant. The white line indicates the scale of 100 μm

mutation of *POLE*. This is a high-grade endometrioid carcinoma with severe nuclear atypia and frequent mitotic figures. Infiltrating lymphocytes are also observed (arrowheads). Although the precise mechanism underlying this favorable prognosis remains unclear, it is expected to be highly immunogenic, possibly related to a good prognosis due to the ultra-mutated genomic characteristics.

4.3.2 MMR-Deficient (MMRd)

MMR is a DNA repair mechanism that fixes errors made during DNA replication, including base mismatches and small insertion or deletion loops. Tumors with intact MMR function are termed MMR-proficient (MMRp), whereas those with defective function are classified as MMR-deficient (MMRd). The MMR mechanism recognizes the mismatch, removes the incorrect section, and fills in the correct DNA sequence using the original strand as a template. This process helps maintain genomic stability and prevent mutations. In MMRd tumors, genomic stability is lost, especially in microsatellite regions. These are short DNA repeat sequences prone to replication errors, leading to variations in microsatellite length and instability, which in turn acquire microsatellite instability (MSI). There are two main methods for clinically detecting MMRd tumors: the MMR-IHC test, which detects the loss of expression of proteins related to the MMR mechanism using IHC, and the MSI test, which detects abnormalities in the repeating sequences of microsatellite regions using PCR or other methods [22, 23].

The MMR-IHC test identifies MMRd as the loss of expression of any of the four MMR proteins on IHC: MLH1, mutL homolog 1; MSH2, mutS homolog 2; MSH6,

Table 4.2 Gene alterations and expression patterns responsible for MMR-IHC

		Expression pattern in IHC			
		MLH1	MSH2	PMS2	MSH6
Responsible gene	Normal	+	+	+	+
	MLH1	–	+	–	+
	MSH2	+	–	+	–
	PMS2	+	+	–	+
	MSH6	+	+	+	–

Normal staining is indicated by +, and loss of staining is indicated by –
IHC immunohistochemistry, *MLH1* mutL homolog 1, *MSH2* mutS homolog 2, *MSH6* mutS homolog 6, *PMS2* PMS1 homolog 2

mutS homolog 6; and PMS2, PMS1 homolog 2. Because these proteins form heterodimers, their loss follows characteristic patterns [24]. Table 4.2 summarizes the gene alterations and expression patterns responsible for MMR-IHC. The PMS2 protein forms heterodimers only with MLH1, whereas the MLH1 protein forms heterodimers with PMS2 and other proteins. The relationship between MSH2 and MSH6 is similar; MSH6 can also detect the loss of MSH2, but MSH6 can only be detected by MSH6. Therefore, screening with MSH6 and PMS2 alone is acceptable when resources are limited.

In the TCGA integrated analysis of uterine cancer, MSI was used to analyze seven microsatellite regions. Some of these regions contain a two-base repeat sequence. Subsequent studies have revealed that single-base repeats have higher sensitivity and specificity. Thus, the abnormality of single-base repeats in five microsatellite regions is now conventionally used in the MSI test by PCR, categorizing cases as high (MSI-H), low (MSI-L), or stable (MSS). Furthermore, NGS has enabled the evaluation of tens to thousands of microsatellite regions. The concordance rate between MMRd and MSI-H in endometrial cancer is reported to be 94% [22, 23].

MMRd/MSI-H was among the earliest molecular markers to influence treatment decisions. The KEYNOTE-158 trial demonstrated the efficacy of anti-PD-1 antibodies in MMRd/MSI-H cases, notably in a tumor-agnostic manner [25]. Current molecular classifications are mainly based on data from cases treated before the widespread use of immune checkpoint inhibitors (ICI). Over the next decade, MMRd/MSI-H cases will show the most remarkable prognostic improvements. However, first-line systemic chemotherapy now includes ICI regardless of MMR status, as a significant proportion of MMRp cases also respond to ICI [26–29].

Endometrial cancer also serves as a sentinel cancer for Lynch syndrome, a hereditary disorder caused by germline pathogenic variants in MMR or EPCAM genes. While all Lynch syndrome cases exhibit MMRd, most MMRd endometrial cancers arise from MLH1 promoter methylation, meaning only about 10% of MMRd cases are attributable to Lynch syndrome [30–33].

4.3.3 p53 Abnormal (p53abn)

TP53 mutations were concentrated in the CN high cluster in TCGA, with 90% of the cluster containing *TP53* mutations. Validation of *TP53* mutations and IHC for the p53 protein has not necessarily been established in endometrial cancer. However, if careful evaluation is performed on p53-IHC, it can be a good surrogate marker for *TP53* mutations. The concordance rate between abnormal p53 IHC and *TP53* mutation is reported as ranging from 65.3% [34] to 91.2% [35]. The accumulation of abnormal p53 is detected as diffuse (>80%) and intense nuclear staining in tumor cells, resulting from missense mutations. The null pattern, from loss of function mutation, is defined as complete expression loss in tumor cells with retained internal control. Rarely, some tumors show a cytoplasmic pattern as an abnormal expression [36]. Abnormal p53 expression is one of the definitive features of morphologic serous carcinoma, but only about 80% of p53 abnormal expression tumors exhibit this morphology, making morphological screening inappropriate [37]. In fact, a study on fertility-sparing treatment cases found that 11.8% had p53 abnormal expression, and their relapse rates at 1 and 2 years were significantly higher than those of p53 wild-type cases (1 year 33.3% versus 15.4%, $P = 0.047$; 2 years 100% versus 40%, $P = 0.036$) [38]. *TP53* is also carcinosarcoma's most frequently mutated gene [39, 40].

Abnormal p53 expression has been reported as a poor prognostic factor in various cohorts. These include cohorts classified based on morphological classification [41], early-stage low-grade endometrioid carcinoma cohorts that do not require adjuvant therapy [42, 43], cohorts where postoperative radiotherapy was primarily chosen as adjuvant therapy [44], and cohorts where cytotoxic chemotherapy was predominantly selected [10]. Consequently, more intensive treatment is recommended [35, 45]. However, the only established treatment options are systemic chemotherapy, external beam radiation therapy, and vaginal brachytherapy, leading to inevitable combinations of these modalities.

4.3.4 No Specific Molecular Profile (NSMP)

NSMP endometrial cancer corresponds to the CN low cluster in TCGA and is negative for all the above molecular markers. Its molecular genomic features are low TMB and SCNAs; its prognosis is intermediate to good. NSMP endometrial cancer does not have any of the abnormalities shown in 3–1 to 3–3, so there is currently no specific treatment available. Stelloo et al. reported that L1CAM has further stratified the prognosis of NSMP, and the prognosis of L1CAM-positive patients was worse than that of negative patients [46]. It is considered a molecular classification that requires further consideration.

4.4 Other Molecular Prognostic Markers

4.4.1 L1CAM

L1CAM is a transmembrane glycoprotein belonging to the immunoglobulin superfamily and plays an important role in the development of the nervous system [47]. Diffusely membranous positive L1CAM staining is a typical histopathological feature of well-differentiated papillary mesothelial tumors, resulting from tumor necrosis factor receptor-associated factor (TRAF) 7 mutations [48]. L1CAM expression has been shown to induce IL-1β production and activate NF-κB through various mechanisms, contributing to invasiveness and metastatic potential in multiple carcinomas, resulting in poor prognosis [49–53]. In endometrial cancer, L1CAM has emerged as an important biomarker associated with high tumor grade and poor prognosis, as confirmed by multiple cohorts [10, 54–62]. It also overlaps with other poor prognostic factors, with more than half of p53abn tumors and a significant proportion of clear cell carcinomas exhibiting high L1CAM expression [10, 54, 56, 60, 62, 63]. However, no standardized method exists for evaluating L1CAM positivity. A large study on FIGO2008 stage I endometrial cancer (n = 1021) proposed an L1CAM positivity threshold of >10% of tumor cells with L1CAM staining [59]. Subsequent reports have confirmed the ability to stratify prognosis using this threshold in other cohorts [55, 56, 58, 62]. However, due to the heterogeneous nature of L1CAM expression, some studies suggest a > 50% threshold [54, 60], while others, including our research, have used the H-score for evaluation [64]. Additionally, TCGA database analysis utilizing RNA-Seq by Expectation-Maximization has also been reported [65].

4.4.2 *CTNNB1*

CTNNB1 mutations hold clinical significance as poor prognostic factors in endometrial cancer, particularly in low-grade carcinoma and NSMP cases, who have no prognostic molecules [62, 63, 66]. *CTNNB1* encodes β-catenin, which is involved in the activation of the Wnt/β-catenin pathway, and abnormalities in this pathway contribute to tumorigenesis and progression [67]. In vivo studies have demonstrated that inhibitors of the Wnt/β-catenin pathway can suppress the proliferation of *CTNNB1*-mutant endometrial cancer cells [68]. Prognostic assessment in endometrial cancer patients suggests that β-catenin IHC can serve as a sufficient surrogate for evaluating *CTNNB1* mutations [69, 70].

4.4.3 HER2 (*ERBB2*)

Human epidermal growth factor receptor 2 (HER2) may serve as an additional treatment target in this high-risk group, as HER2 overexpression and/or amplification is observed in 35% of serous carcinoma [71], 23–27% of clear cell carcinoma [72], and 9% of carcinosarcoma [39, 40]. A phase II randomized trial evaluating the addition of the anti-HER2 antibody trastuzumab to conventional carboplatin/paclitaxel therapy in advanced/recurrent HER2-positive serous carcinoma confirmed an improvement in overall survival (median 29.6 months versus 24.4 months; hazard ratio = 0.58; 90% confidence interval, 0.34–0.99; $P = 0.046$) [73]. In addition, the DESTINY PanTumor02 study, which investigated the efficacy of trastuzumab deruxtecan (T-DXd), an antibody-drug conjugate made by combining trastuzumab with a topoisomerase I inhibitor in patients with HER2-positive advanced recurrent solid tumors, showed a primary endpoint of 37.1% objective response rate (ORR), with an ORR of 57.5% in patients with uterine cancer ($N = 40$), and an exceptionally favorable result of 84.6% in patients with 3+ IHC ($N = 13$) [74]. Furthermore, the STATICE study, which evaluated the efficacy and safety of T-DXd to recurrent uterine carcinosarcoma with HER2 evaluation of 1+ or higher by IHC, showed the ORR for 22 HER2-high and 10 HER2-low patients was 54.4% (95%CI, 32.2–75.6) and 70.0% (95%CI, 34.8–93.3), respectively, and the median progression-free survival was 6.2 and 6.7 months, respectively [75]. Testing for HER2 includes IHC for HER2 or amplification of ERBB2 in tissue or circulating cell-free DNA using NGS [74–76], and HER2 testing should be considered in p53 abnormal expression tumors regardless of histologic types.

4.5 Updates of FIGO Staging System

4.5.1 Overview of Update Points of FIGO2023

The primary revision of the 2023 FIGO classification is that it is no longer a simple classification of progression. The FIGO 2023 staging system is the first to include molecular genetic prognostic factors in staging endometrial cancer. It also reflects the latest research on pathologic risk factors for recurrence and/or lymph node metastasis. The classification has become highly complex, increasing from 9 categories in the FIGO 2008 staging system to 19 categories plus two molecular categories, more than doubling.

In stages I to II, molecular classification, histological type, and degree of LVSI are now considered in addition to the simple progression classification, summarized in Fig. 4.5a. Molecular classification is the most important prognostic factor in stage I to II endometrial cancer. Thus, *POLE*mut is subclassified as $\mathrm{IAm}_{\mathit{POLE}\mathrm{mut}}$ and p53abn as $\mathrm{IICm}_{\mathrm{p53abn}}$, regardless of histological type, LVSI, or tumor progression. The histological subtype is the next important factor and divided into non-aggressive and aggressive types. Particularly in non-aggressive types, it is essential to note that substage changes depending on the progression of the tumor and the degree of LVSI. When LVSI is no or focal in non-aggressive histological type, tumors limited to the endometrium, tumors that invade less than half of the myometrium, tumors that invade half or more of the myometrium, tumors with cervical stroma invasion, and tumors that invade one side of the ovary without invasion of the myometrium are classified as IA1, IA2, IB, IIA, and IA3, respectively. However, when LVSI is

	Stage I Confined to the uterine corpus and ovary*					Stage II Invasion of cervical stroma without extrauterine extension OR with substyantial LVSI OR aggressive histological types with myometrial invasion	
Disease limited to	endometrial polyp OR endometrium	less than half of myometrium	half or more of myometrium	less than half of myometrium	half or more of myometrium	cervical stroma	
LVSI	no	no or focal		substantial		no or focal	substantial
LNM	no						
Non-aggressive histological type	I A1	I A2	I B	II B		II A	II B
Aggressive histological types	I C	II C					
POLE mut	I A1 m *POLE*mut						
p53abn	II Cm p53abn						

	Stage III Local and/or regional spread of the tumor of any histologicals subtype									Stage IV Spread to the bladder mucosa and/or intestinal mucosa and/or distance metastasis		
Disease limited to	ovary or fallopian tube		uterine subserosa or spread through the uterine serosa	vagina and/or the parametria	pelvic peritoneum	IIIC1: pelvic lymph nodes		IIIC2 Metastasis to para- aortic lymph nodes up to the renal vessels		Invasion of the bladder mucosa and/or the intestinal/bowel mucosa	Abdominal peritoneal metastasis beyond the pelvis	Distant metastasis, including metastasis to any extra- or intra-abdominal lymph nodes above the renal vessels, lungs, liver, brain, or bone
LVSI	no or focal	sub-stantial	any									
LNM	no					Micro	Macro	Micro	Macro	any		
Non-aggressive histological type	IA3*	IIIA1	IIIA2	IIIB1	IIIB2	IIIC1i	IIIC1ii	IIIC2i	IIIC2ii	IVA	IVB	IVC
Aggressive histological type	IIIA1											
POLEmut												
p53abn												

Fig. 4.5 FIGO2023 staging system for endometrial cancer. FIGO2023 staging system for stage I and II (**a**) and stage III and IV (**b**) endometrial cancer. Non-aggressive histological type is low-grade (grade 1 or 2) endometrioid carcinoma. *See text for distinction between stage IA3 and stage IIIA1. LVSI lymphovascular space invasion, LNM lymph node metastasis, *POLE*mut DNA polymerase epsilon mutation, p53abn p53 abnormal

substantial in non-aggressive histological type, it is classified as IIB regardless of the depth of myometrium invasion or invasion to the cervical stroma. On the other hand, in aggressive histological type, when the tumor is confined to the endometrium, it is classified as stage IC; otherwise, it is classified as stage IIC, regardless of the degree of myometrial invasion or LVSI.

In stages III to IV, the essential revisions are the more detailed substage of tumor spread and the distinction between micro and macro lymph node metastases, shown in Fig. 4.5b. For example, stage IIIA1 with adnexal metastasis is distinguished from stage IIIA2 with serosal infiltration, and peritoneal metastasis/dissemination is classified as stage IIIB2 if confined to the pelvis or as stage IVB if it extends beyond the pelvis. Here, we summarize the evaluation methods for each factor below.

4.5.2 Molecular Classification in FIGO2023

Assessments of *POLE*, MMRd/MSI, and p53 were encouraged in all patients with endometrial cancer. In stages I/II, regardless of histological type and vascular invasion, *POLE*mut is classified separately as $\mathrm{IAm}_{POLE\mathrm{mut}}$, and p53abn is classified as $\mathrm{IICm}_{\mathrm{p53abn}}$. *POLE*mut may show a highly atypical (aggressive) morphology; thus, if *POLE* evaluation is not performed, more cases may be classified as stage IC/IIC. In stages III/IV, molecular status does not change the stage but is noted by adding "m" after the stage to indicate the molecular classification. These molecular assessments can be conducted using either biopsy or hysterectomy specimens, but the specific modality is not defined.

4.5.3 Histological Subtypes and Grade

The only non-aggressive histological type is endometrioid carcinoma grade 1/2. In low-grade tumors, stage IA3 with endometrial and ovarian involvement is notable, as it was mentioned that adjuvant therapy is unnecessary. However, distinguishing it from stage IIIA1 is crucial. To be classified as stage IA3, all the following criteria must be met: (1) no more than superficial myometrial invasion (<50%); (2) absence of extensive/substantial LVSI; (3) absence of additional metastases; and (4) the ovarian tumor is unilateral, limited to the ovary, without capsule invasion/rupture (equivalent to pT1a). For aggressive histological types, even if the tumor is limited to a polyp or lacks myometrial invasion, it is classified as stage IC. If myometrial invasion is present, it is classified as stage IIC or higher.

4.5.4 Lymphovascular Space Invasion (LVSI)

Instead of merely assessing the presence of vascular invasion, the extent of invasion is now included. Whether vascular invasion is a prognostic factor and whether its evaluation is consistent among assessors have long been debated. The new FIGO classification system distinguishes between "no or focal" and "extensive/substantial." The threshold is five vessels, based on the WHO Classification of Tumours fifth ed. recommendations [7]. However, this threshold is controversial. The sensitivity of LVSI detection depends on the evaluation method. A key factor is whether to use only H&E-stained sections or add immunohistochemical stains. Widely used immunohistochemical markers for LVSI assessment include D2–40 (podoplanin, a lymphatic vessel marker) and CD31 (a venous marker), which are highly useful for detecting small or discontinuous lesions. Additionally, the number and location of sections assessed affect detection sensitivity, and following the recommendations of the International Society of Gynecologic Pathologists (ISGYP) is advised [77]. These factors are thought to have little impact on evaluating the extent of vascular invasion [78, 79]. Thus, in assessing the extent rather than presence, using H&E-stained slides from the deepest invasive area is sufficient and reduces interobserver discrepancies.

Mimic LVSI due to MELF-type invasion was also mentioned. MELF (microcystic, elongated, and fragmented) is an invasion pattern described in the WHO Classification of Tumours fifth ed. that is associated with deep myometrial invasion and lymph node metastasis [80]. Still, it is not linked to prognosis [81]. Flattened tumor cells surrounding vessels may be misinterpreted as LVSI. Tumor cells can be distinguished from actual endothelial cells by cytokeratin immunostaining.

LVSI assessment is primarily based on hysterectomy specimens, making evaluation difficult in cases treated with fertility-preserving endocrine therapy or radiation therapy due to complications.

4.5.5 Lymph Node Metastasis

Lymph node metastasis assessment now includes microscopic evaluation of metastatic foci size. The criteria are macro- >2 mm, micro- 0.2–2 mm and/or > 200 cells, and isolated tumor cells (ITCs) <0.2 mm and < 200 cells [82]. Micrometastasis or larger is considered positive, with micro- classified as "i" and macro- as "ii" within IIIC1/2. ITCs are deemed negative, but TNM classification records them as pN0(i+) [82]. These updates are based on sentinel lymph node biopsy (SLNB) and ultrastaging. NCCN guidelines state that SLNB "can be considered for patients with apparent uterine-confined disease" [45]. ESGO/ESTRO/ESP guidelines recommend SLNB for low−/intermediate-risk cases with myometrial invasion and as an acceptable alternative to systemic lymphadenectomy for high−/high-intermediate-risk cases [35]. When performing sentinel lymph node mapping, SLN should be evaluated using ultrastaging. Although ultrastaging methods are not strictly defined, serial sectioning with H&E staining, with or without pancytokeratin IHC staining, is commonly accepted [35, 45, 83–86]. Ultrastaging is intended for SLN specimens, which are few in number. In early-stage endometrial cancer, metastases ≤2 mm occur in 1.9–10.2% of cases, comprising 25–63% of all metastases, and ultrastaging reduces false negatives from 25 to 48% [84–89] (see Chap. 7).

4.6 Future Directions

The new FIGO staging system has evolved from a simple stage classification to an integrated system incorporating both morphological and molecular genetic assessments. This enhancement improves prognostic stratification, but standardizing evaluation methods for international comparison remains crucial. Molecular subtyping enables treatment de-escalation for favorable prognosis patients and intensified treatment for poor prognosis patients. Furthermore, when molecular markers serve as predictive biomarkers, personalized therapy becomes feasible. The PORTEC study group is investigating whether *POLE*mut cases, despite being classified as conventional high-intermediate risk, can safely omit postoperative radiotherapy [16]. In advanced disease, immune checkpoint inhibitors will be incorporated into first-line treatment, potentially improving outcomes, particularly in MMRd cases [26–29]. While treatment intensification has been proposed for patients with p53 abnormal expression and who are L1CAM positive, further development of targeted therapies is anticipated [35, 45].

References

1. Bokhman JV. Two pathogenetic types of endometrial carcinoma. Gynecol Oncol. 1983;15(1):10–7.
2. Cancer Genome Atlas Research, N, et al. Integrated genomic characterization of endometrial carcinoma. Nature. 2013;497(7447):67–73.
3. Berek JS, et al. FIGO staging of endometrial cancer: 2023. Int J Gynaecol Obstet. 2023;162(2):383–94.
4. The National Cancer Institute, t.N.I.o.H., U.S. Department of Health and Human Services. The cancer genome atlas program. 2024 [cited 2025 January 12]. Available from: https://www.cancer.gov/ccg/research/genome-sequencing/tcga.
5. Stelloo E, et al. Refining prognosis and identifying targetable pathways for high-risk endometrial cancer; a TransPORTEC initiative. Mod Pathol. 2015;28(6):836–44.
6. Talhouk A, et al. A clinically applicable molecular-based classification for endometrial cancers. Br J Cancer. 2015;113(2):299–310.
7. Board., W.C.o.T.E. Female genital tumours, WHO classification of tumours, vol. 4. 5th ed. IARC Press; 2020.
8. Hammer PM, et al. Molecular classification outperforms histologic classification in prognostication of high-grade endometrial carcinomas with undifferentiated and sarcomatous components. Am J Surg Pathol. 2024;48(8):953–64.
9. Kim SR, et al. Molecular subtypes of clear cell carcinoma of the endometrium: opportunities for prognostic and predictive stratification. Gynecol Oncol. 2020;158(1):3–11.
10. Yamazaki H, et al. The prognosis of endometrial cancers stratified with conventional risk factors and modified molecular classification. Cancer Sci. 2022;113:3134.
11. Asami Y, et al. Utility of molecular subtypes and genetic alterations for evaluating clinical outcomes in 1029 patients with endometrial cancer. Br J Cancer. 2023;128(8):1582–91.
12. McAlpine JN, et al. Evaluation of treatment effects in patients with endometrial cancer and POLE mutations: an individual patient data meta-analysis. Cancer. 2021;127(14):2409–22.
13. Jumaah AS, et al. The frequency of POLE-mutation in endometrial carcinoma and prognostic implications: a systemic review and meta-analysis. J Pathol Transl Med. 2020;54(6):471–9.
14. Stasenko M, et al. Clinical outcomes of patients with POLE mutated endometrioid endometrial cancer. Gynecol Oncol. 2020;156(1):194–202.
15. Horeweg N, et al. Molecular classification predicts response to radiotherapy in the randomized PORTEC-1 and PORTEC-2 trials for early-stage endometrioid endometrial cancer. J Clin Oncol. 2023;41(27):4369–80.
16. van den Heerik A, et al. PORTEC-4a: international randomized trial of molecular profile-based adjuvant treatment for women with high-intermediate risk endometrial cancer. Int J Gynecol Cancer. 2020;30(12):2002–7.
17. Church DN, et al. Prognostic significance of POLE proofreading mutations in endometrial cancer. J Natl Cancer Inst. 2015;107(1):402.

18. Sondka Z, et al. COSMIC: a curated database of somatic variants and clinical data for cancer. Nucleic Acids Res. 2024;52(D1):D1210–7.
19. Leon-Castillo A, et al. Interpretation of somatic POLE mutations in endometrial carcinoma. J Pathol. 2020;250(3):323–35.
20. The cBioPortal for Cancer Genomics. [cited 2025 January 3]. Available from: https://www.cbioportal.org/.
21. Yanazume S, et al. Cytopathological features associated with POLE mutation in endometrial cancer. Cytopathology. 2023;34(3):211.
22. Stelloo E, et al. Practical guidance for mismatch repair-deficiency testing in endometrial cancer. Ann Oncol. 2017;28(1):96–102.
23. Schöniger S, Rüschoff J. Mismatch repair deficiency and microsatellite instability. Encyclopedia. 2022;2(3):1559–76.
24. Shia J, et al. Immunohistochemistry as first-line screening for detecting colorectal cancer patients at risk for hereditary nonpolyposis colorectal cancer syndrome: a 2-antibody panel may be as predictive as a 4-antibody panel. Am J Surg Pathol. 2009;33(11):1639–45.
25. Marabelle A, et al. Efficacy of pembrolizumab in patients with noncolorectal high microsatellite instability/mismatch repair-deficient cancer: results from the phase II KEYNOTE-158 study. J Clin Oncol. 2020;38(1):1–10.
26. Eskander RN, et al. Pembrolizumab plus chemotherapy in advanced endometrial cancer. N Engl J Med. 2023;388(23):2159–70.
27. Van Gorp T, et al. ENGOT-en11/GOG-3053/KEYNOTE-B21: a randomised, double-blind, phase III study of pembrolizumab or placebo plus adjuvant chemotherapy with or without radiotherapy in patients with newly diagnosed, high-risk endometrial cancer. Ann Oncol. 2024;35(11):968–80.
28. Baurain J-F, et al. Durvalumab plus carboplatin/paclitaxel followed by durvalumab with or without olaparib as a firstline treatment for endometrial cancer: overall survival and additional secondary efficacy en-dpoints by mismatch repair status in the DUO-E/GOG-3041/ENGOT-EN10 trial. Gynecol Oncol. 2024;190:S62–3.
29. Powell MA, et al. Overall survival in patients with endometrial cancer treated with dostarlimab plus carboplatin-paclitaxel in the randomized ENGOT-EN6/GOG-3031/RUBY trial. Ann Oncol. 2024;35(8):728–38.
30. Porkka N, et al. Epidemiological, clinical and molecular characterization of Lynch-like syndrome: a population-based study. Int J Cancer. 2019;145(1):87–98.
31. Hampel H, et al. Double somatic mismatch repair gene pathogenic variants as common as Lynch syndrome among endometrial cancer patients. Gynecol Oncol. 2021;160(1):161–8.
32. Post CCB, et al. Prevalence and prognosis of Lynch syndrome and sporadic mismatch repair deficiency in endometrial cancer. J Natl Cancer Inst. 2021;113(9):1212–20.
33. Kato MK, et al. Clinical features and impact of p53 status on sporadic mismatch repair deficiency and Lynch syndrome in uterine cancer. Cancer Sci. 2024;115(5):1646–55.
34. Guo Q, et al. Identification of molecular subtypes for endometrial carcinoma using a 46-gene next-generation sequencing panel: a retrospective study on a consecutive cohort. ESMO Open. 2024;9(10):103710.
35. Concin N, et al. ESGO/ESTRO/ESP guidelines for the management of patients with endometrial carcinoma. Int J Gynecol Cancer. 2021;31(1):12–39.
36. Kobel M, et al. Interpretation of P53 immunohistochemistry in endometrial carcinomas: toward increased reproducibility. Int J Gynecol Pathol. 2019;38 Suppl 1(Iss 1 Suppl 1):S123–31.
37. Vaziri Fard E, et al. p53 abnormal (copy number high) endometrioid endometrial carcinoma has a prognosis indistinguishable from serous carcinoma. Int J Gynecol Pathol. 2024;43(5):515–26.
38. Peng H, et al. Mismatch repair deficiency and abnormal p53 expression has significant predictive value for progesterone resistance and endometrial tumorigenesis in patients with endometrial atypical hyperplasia receiving fertility-preserving treatment. Gynecol Oncol. 2024;186:154–60.

39. Moukarzel LA, et al. The genetic landscape of metaplastic breast cancers and uterine carcinosarcomas. Mol Oncol. 2021;15(4):1024–39.
40. Crane E, et al. Molecular variations in uterine carcinosarcomas identify therapeutic opportunities. Int J Gynecol Cancer. 2020;30(4):480–4.
41. Ito K, et al. Prognostic significance of p53 overexpression in endometrial cancer. Cancer Res. 1994;54(17):4667–70.
42. Chang YW, et al. Abnormal p53 expression is associated with poor outcomes in grade I or II, stage I, endometrioid carcinoma: a retrospective single-institute study. J Gynecol Oncol. 2024;35(6):e78.
43. Puppo A, et al. Where morphological and molecular classifications meet: the role of p53 immunohistochemistry in the prognosis of low-risk endometrial carcinoma (GLAMOUR study). Cancers (Basel). 2024;16(6):1088.
44. Raffone A, et al. TCGA molecular groups of endometrial cancer: pooled data about prognosis. Gynecol Oncol. 2019;155(2):374–83.
45. Network, N.C.C. Uterine neoplasms (version 2.2025). 2025 January 31, 2025 [cited 2025 February 8]; 2.2025. Available from: https://www.nccn.org/professionals/physician_gls/pdf/uterine.pdf.
46. Stelloo E, et al. Improved risk assessment by integrating molecular and clinicopathological factors in early-stage endometrial cancer-combined analysis of the PORTEC cohorts. Clin Cancer Res. 2016;22(16):4215–24.
47. Hortsch M. Structural and functional evolution of the L1 family: are four adhesion molecules better than one? Mol Cell Neurosci. 2000;15(1):1–10.
48. Goode B, et al. Adenomatoid tumors of the male and female genital tract are defined by TRAF7 mutations that drive aberrant NF-kB pathway activation. Mod Pathol. 2018;31(4):660–73.
49. Kiefel H, et al. L1CAM: a major driver for tumor cell invasion and motility. Cell Adhes Migr. 2012;6(4):374–84.
50. Altevogt P, Doberstein K, Fogel M. L1CAM in human cancer. Int J Cancer. 2016;138(7):1565–76.
51. Kiefel H, et al. Linking L1CAM-mediated signaling to NF-kappaB activation. Trends Mol Med. 2011;17(4):178–87.
52. Kiefel H, et al. EMT-associated up-regulation of L1CAM provides insights into L1CAM-mediated integrin signalling and NF-kappaB activation. Carcinogenesis. 2012;33(10):1919–29.
53. Kiefel H, et al. L1CAM-integrin interaction induces constitutive NF-kappaB activation in pancreatic adenocarcinoma cells by enhancing IL-1beta expression. Oncogene. 2010;29(34):4766–78.
54. Van Gool IC, et al. Prognostic significance of L1CAM expression and its association with mutant p53 expression in high-risk endometrial cancer. Mod Pathol. 2016;29(2):174–81.
55. Bosse T, et al. L1 cell adhesion molecule is a strong predictor for distant recurrence and overall survival in early stage endometrial cancer: pooled PORTEC trial results. Eur J Cancer. 2014;50(15):2602–10.
56. Kommoss FK, et al. L1CAM further stratifies endometrial carcinoma patients with no specific molecular risk profile. Br J Cancer. 2018;119(4):480–6.
57. Fogel M, et al. L1 expression as a predictor of progression and survival in patients with uterine and ovarian carcinomas. Lancet. 2003;362(9387):869–75.
58. van der Putten LJ, et al. L1CAM expression in endometrial carcinomas: an ENITEC collaboration study. Br J Cancer. 2016;115(6):716–24.
59. Zeimet AG, et al. L1CAM in early-stage type I endometrial cancer: results of a large multicenter evaluation. J Natl Cancer Inst. 2013;105(15):1142–50.
60. Fadare O, et al. The significance of L1CAM expression in clear cell carcinoma of the endometrium. Histopathology. 2018;72(3):532–8.
61. Hua T, et al. Prognostic significance of L1 cell adhesion molecule in cancer patients: a systematic review and meta-analysis. Oncotarget. 2016;7(51):85196–207.
62. Kim J, et al. Prognostic significance of L1CAM expression in addition to ProMisE in endometrial cancer. Gynecol Oncol. 2023;174:231–8.

63. Yoon H, et al. Evaluation of prognostic potential of beta-catenin and L1CAM expression according to endometrial cancer risk group. Gynecol Oncol. 2024;184:132–8.
64. Asano H, et al. L1CAM predicts adverse outcomes in patients with endometrial cancer undergoing full lymphadenectomy and adjuvant chemotherapy. Ann Surg Oncol. 2020;27(7):2159–68.
65. Dellinger TH, et al. L1CAM is an independent predictor of poor survival in endometrial cancer – an analysis of the cancer genome atlas (TCGA). Gynecol Oncol. 2016;141(2):336–40.
66. Travaglino A, et al. Prognostic significance of CTNNB1 mutation in early stage endometrial carcinoma: a systematic review and meta-analysis. Arch Gynecol Obstet. 2022;306(2):423–31.
67. Ledinek Z, Sobocan M, Knez J. The role of CTNNB1 in endometrial cancer. Dis Markers. 2022;2022:1442441.
68. Moroney MR, et al. Inhibiting Wnt/beta-catenin in CTNNB1-mutated endometrial cancer. Mol Carcinog. 2021;60(8):511–23.
69. Travaglino A, et al. Immunohistochemical nuclear expression of beta-catenin as a surrogate of CTNNB1 exon 3 mutation in endometrial cancer. Am J Clin Pathol. 2019;151(5):529–38.
70. Kim G, et al. Nuclear beta-catenin localization and mutation of the CTNNB1 gene: a context-dependent association. Mod Pathol. 2018;31(10):1553–9.
71. Buza N, et al. Toward standard HER2 testing of endometrial serous carcinoma: 4-year experience at a large academic center and recommendations for clinical practice. Mod Pathol. 2013;26(12):1605–12.
72. Sukov WR, et al. Frequency of HER2 protein overexpression and HER2 gene amplification in endometrial clear cell carcinoma. Hum Pathol. 2023;137:94–101.
73. Fader AN, et al. Randomized phase II trial of carboplatin-paclitaxel compared with carboplatin-paclitaxel-trastuzumab in advanced (stage III-IV) or recurrent uterine serous carcinomas that overexpress Her2/Neu (NCT01367002): updated overall survival analysis. Clin Cancer Res. 2020;26(15):3928–35.
74. Meric-Bernstam F, et al. Efficacy and safety of trastuzumab deruxtecan in patients with HER2-expressing solid tumors: primary results from the DESTINY-PanTumor02 phase II trial. J Clin Oncol. 2024;42(1):47–58.
75. Nishikawa T, et al. Trastuzumab deruxtecan for human epidermal growth factor receptor 2-expressing advanced or recurrent uterine carcinosarcoma (NCCH1615): the STATICE trial. J Clin Oncol. 2023;41(15):2789–99.
76. Yagisawa M, et al. Trastuzumab deruxtecan in advanced solid tumors with human epidermal growth factor receptor 2 amplification identified by plasma cell-free DNA testing: a multicenter, single-arm, phase II basket trial. J Clin Oncol. 2024;42(32):3817–25.
77. Malpica A, et al. Endometrial carcinoma, grossing and processing issues: recommendations of the international society of gynecologic pathologists. Int J Gynecol Pathol. 2019;38(Suppl 1(Iss 1 Suppl 1)):S9–S24.
78. Peters EEM, et al. Reproducibility of lymphovascular space invasion (LVSI) assessment in endometrial cancer. Histopathology. 2019;75(1):128–36.
79. Peters EEM, et al. Practical guidance for assessing and reporting lymphovascular space invasion (LVSI) in endometrial carcinoma. Histopathology. 2025;86(2):173–82.
80. Pavlakis K, et al. MELF invasion in endometrial cancer as a risk factor for lymph node metastasis. Histopathology. 2011;58(6):966–73.
81. van den Heerik A, et al. Microcystic elongated and fragmented (MELF) pattern of invasion: molecular features and prognostic significance in the PORTEC-1 and -2 trials. Gynecol Oncol. 2022;166(3):530–7.
82. Control, U.f.I.C. TNM classification of malignant tumours. 8th ed. Wiley-Blackwell; 2016.
83. Cormier B, et al. Sentinel lymph node procedure in endometrial cancer: a systematic review and proposal for standardization of future research. Gynecol Oncol. 2015;138(2):478–85.
84. Euscher E, et al. Ultrastaging of sentinel lymph nodes in endometrial carcinoma according to use of 2 different methods. Int J Gynecol Pathol. 2018;37(3):242–51.
85. Lavecchia M, et al. Sentinel lymph node biopsy in endometrial cancer: the new norm – a multicentre, international experience. Surg Oncol. 2023;48:101922.

86. Mueller JJ, et al. Incidence of pelvic lymph node metastasis using modern FIGO staging and sentinel lymph node mapping with ultrastaging in surgically staged patients with endometrioid and serous endometrial carcinoma. Gynecol Oncol. 2020;157(3):619–23.
87. Delpech Y, et al. Sentinel lymph node evaluation in endometrial cancer and the importance of micrometastases. Surg Oncol. 2008;17(3):237–45.
88. Kim CH, et al. Sentinel lymph node mapping with pathologic ultrastaging: a valuable tool for assessing nodal metastasis in low-grade endometrial cancer with superficial myoinvasion. Gynecol Oncol. 2013;131(3):714–9.
89. Buda A, et al. The risk of recurrence in endometrial cancer patients with low-volume metastasis in the sentinel lymph nodes: a retrospective multi-institutional study. Cancers (Basel). 2023;15(7):2052.

Artificial Intelligence in the Diagnosis of Endometrial Cancer

5

Yusuke Toyohara and Kenbun Sone

Abstract

Artificial intelligence (AI) aims to replicate human cognitive abilities, such as problem-solving and decision-making. Early AI relied on logical reasoning and rule-based systems, but advancements led to machine learning (ML), where models learn from data to make predictions. Deep learning (DL), a subset of ML, addresses the challenges of handling unstructured data by utilizing artificial neural networks (ANNs), particularly deep neural networks (DNNs). AI is transforming medical diagnostics, especially in endoscopy, imaging (X-rays, MRI, CT scans), and waveform analysis (ECGs). In MRI-based preoperative diagnosis of endometrial cancer, AI enhances tumor detection, myometrial invasion assessment, and segmentation accuracy. In some cases, AI models even outperform radiologists. Additionally, AI-assisted pathology using hematoxylin and eosin (H&E)-stained slides improves histopathological diagnosis and molecular classification of endometrial cancer, aiding treatment decisions. AI-based hysteroscopy is also enhancing diagnostic accuracy. As AI continues to evolve, its integration into medical imaging, pathology, and real-time diagnostics is revolutionizing healthcare and cancer detection.

Keywords

Artificial intelligence · Deep learning · Molecular subtype · Hysteroscopy

Y. Toyohara · K. Sone (✉)
The University of Tokyo, Bunkyo, Japan

H. Watari, T. Mitamura (eds.), *Uterine Body Neoplasms*, Comprehensive Gynecology and Obstetrics, https://doi.org/10.1007/978-981-95-0946-1_5

5.1 Overview of AI, Machine Learning, and Deep Learning

Artificial intelligence (AI) refers to technologies and concepts that enable computers to mimic intelligent behavior. AI research primarily aims to replicate human cognitive abilities, such as problem-solving, decision-making, learning, and pattern recognition, to automate a wide range of tasks. Early AI, also known as classical AI, was established in the 1950s and the 1960s and relied on techniques such as logical reasoning and rule-based systems. Examples include Chess programs and early expert systems [1]. As AI progressed, it evolved into machine learning (ML), which focuses on building models that learn from data to make predictions. ML can be categorized into three main learning methods: (1) Supervised learning: Models learn from input data and the corresponding labeled outputs. It is commonly used for classification and regression tasks. (2) Unsupervised learning: Models analyze unlabeled data to uncover structures or patterns, with clustering and dimensionality reduction as notable examples. (3) Reinforcement learning: AI agents interact with their environments to learn optimal actions by maximizing rewards through trial and error [2]. Before the advent of deep learning (DL), machine learning faced challenges in processing unstructured data, such as images, audio, and text, requiring manual feature extraction. To address these limitations, deep learning has emerged as a subset of machine learning, which employs artificial neural networks (ANNs) inspired from the human brain (Fig. 5.1). Deep neural networks (DNNs) with multilayer structures automatically extract features and perform tasks with high accuracy. This has led to significant advancements in fields such as image recognition, speech recognition, and natural language processing and has become a central part of the AI revolution [3] (Fig. 5.2). The success of deep learning is attributed to several factors:

Availability of big data: The proliferation of large datasets facilitates model training. Advancements in computational resources: High-performance hardware,

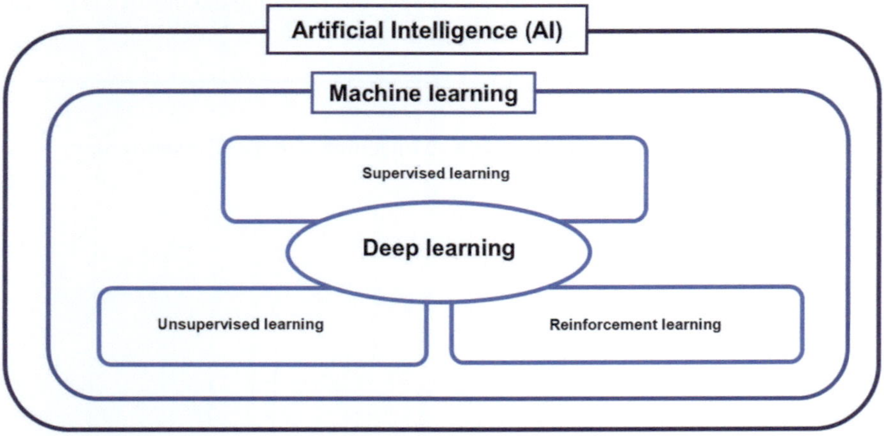

Fig. 5.1 Diagram of artificial intelligence, machine learning, and deep learning

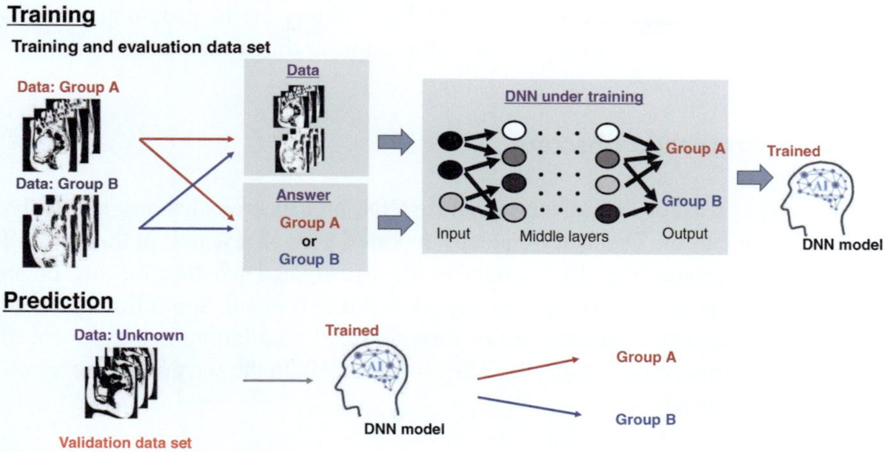

Fig. 5.2 Example of deep learning model development

such as GPUs and TPUs, enable efficient training of large-scale networks. Improved algorithms: Convolutional neural networks (CNNs) are used for image recognition, whereas recurrent neural networks (RNNs) perform well with sequential data. AI has found practical applications across industries including finance, transportation, and entertainment and is increasingly being adopted in the medical field. However, challenges such as the "black box" nature of deep-learning models, high computational resource demands, and ethical concerns remain significant hurdles [4].

5.2 AI and Medical Devices (Including Other Cancers)

AI research has expanded into the medical field, leading to the approval of AI-powered medical devices worldwide. The latest AI medical devices function as diagnostic tools and can be broadly categorized as follows: (1) endoscopy diagnostics: systems such as those for gastrointestinal endoscopy or colonoscopy, which identify lesions in real time during examinations; (2) imaging modalities: AI tools that analyze X-rays, CT scans, and MRI images; (3) waveform analysis: systems analyzing electrical signals, such as ECGs.

5.2.1 Example: Colonoscopy Diagnostics

An artificial intelligence (AI) system was developed to automatically detect early signs of colorectal cancer during colonoscopy, significantly reducing the risk of missed lesions. This system demonstrated a sensitivity of 97.3% (95% CI: 95.9–98.4%) and a specificity of 99.0% (95% CI: 98.6–99.2%) during validation [5]. Further enhancements led to the creation of "WISE VISION Endoscopic Image Analysis AI," a colonoscopy diagnostic support software. During the examination,

AI highlights suspicious areas with visual and auditory alerts, enabling real-time collaboration between endoscopists and AI to improve diagnostic accuracy.

5.2.2 Example: Colposcopy

Colposcopy is used to definitively diagnose CIN, precancerous lesions, and early-stage cervical cancer. This is a frequently reported area of research in the fields of obstetrics and gynecology [6–8]. Kim et al. investigated whether the use of an AI-powered colposcopy system, Cerviray AI® (AIDOT, Seoul, South Korea), could improve the diagnostic performance of four physicians, including two experienced and two inexperienced practitioners. They found that both the sensitivity and specificity were enhanced [9].

5.2.3 Example: Dermatology

In dermatology, melanoma accounts for the majority of skin cancer-related deaths; however, diagnosis is often challenging because of its similarity to benign lesions. Using deep learning, a diagnostic AI system for pigmented skin lesions was developed based on 120,000 images, including those of melanomas, basal cell carcinomas, and benign conditions such as seborrheic keratosis. AI achieved a diagnostic accuracy comparable to that of dermatologists and general practitioners [10].

5.2.4 Example: Breast Cancer

ProFound AI Version 2.1 (iCAD), approved by the FDA in October 2019, is a CAD software for digital breast tomosynthesis (DBT). This system assists in the accurate diagnosis of breast cancer by detecting soft-tissue densities, architectural distortions, asymmetries, and calcifications in 3D DBT slices [11].

5.2.5 Example: Cardiology

AliveCor Kardia Mobile, an AI-powered ECG analysis system, operates on smartphones and tablets to detect arrhythmias such as atrial fibrillation (AF). A study by Goldenthal et al. demonstrated that this system increased the likelihood of detecting recurrent arrhythmias compared to a control group (hazard ratio = 1.56, 95% CI: 1.06–2.30, $P = 0.024$) [12].

5.2.6 Example: Cardiotocography

Cardiotocography (CTG) is a diagnostic tool that records fetal heart rate (FHR) and uterine contractions to assess fetal health. Consequently, AI research has been conducted on CTGs. For example, Imane et al. developed a model called "DeepCTG® 1.0" to predict fetal acidosis using CTG data. This model utilizes several features extracted from CTG data. The developed AI model demonstrated a strong performance in predicting UApH (umbilical artery pH) values (<7.05) [13].

5.3 The Use of AI in Preoperative Imaging Diagnosis of Endometrial Cancer

In the preoperative diagnosis of endometrial cancer, MRI is important, especially in patients whose condition cannot be determined by endometrial cytology or histology, and for diagnosing deep myometrial invasion and lymph node metastasis to determine preoperative staging precisely. With dramatic advances in AI, many studies using AI have recently been reported.

Several studies have been conducted to diagnose masses in the uterine cavity. Zhang et al. [14] developed the DL model for the diagnosis of endometrial cancer using the sagittal T1- and T2-weighted images of 158 patients with endometrial cancer. They used radiomics based on a CNN model using LeNet-5. The comprehensive prediction model based on clinicopathological information and imaging parameters showed a high diagnostic rate of AUC 0.913 in the training set and AUC 0.897 in the test set. Urushihara et al. [15] developed another DL model for diagnosing endometrial cancer using T2-weighted images (T2WI), apparent diffusion coefficient (ADC) maps, and contrast-enhanced T1-weighted images (T1WI) for 204 endometrial cancer images and 184 normal uterine images. The results showed that the performance of each MRI sequence was AUC 0.88–0.95, which was significantly higher than that of the three radiologists, especially for the axial ADC map and axial contrast-enhanced T1WI. Interestingly, combining the sequences did not improve the results, with AUCs of 0.87–0.93. ADC values have been reported to be significantly lower in endometrial cancer than in endometrial polyps and the normal myometrium [16], and the DL model may be better able to recognize differences in texture than humans. In fact, the DL model could diagnose small nodules that radiologists could not recognize as endometrial cancer positive in this study.

In addition to investigations on the diagnosis of endometrial cancer with AI, various studies have reported the evaluation of deep myometrial invasion in preoperative MRI using AI because the preoperative evaluation of deep myometrial invasion has a significant impact on the decision for surgical strategies in endometrial cancer. Dong et al. [17] developed DL models conducted by U-Net with ResNet34 and VGG using contrast-enhanced T1WI and T2WI in 72 patients with stage IA and IB endometrial cancer. The accuracies of T1WI and T2WI were 79.2% and 70.8%, respectively, whereas that of the two radiologists was 77.8%. Thus, the results of the

DL models were comparable with those of the radiologists. Interestingly, the DL model showed a low positivity rate in patients with uterine leiomyomas and endometrial polyps, indicating a challenge for the future development of DL models. Tao et al. [18] also constructed a DL model using preoperative MRI images of 80 patients with endometrial cancer and reported an accuracy of 88.75%, a sensitivity of 87.5%, and a specificity of 95% for the evaluation of deep myometrial invasion. In the development of DL models, tumor segmentation is important for the improvement of precise diagnosis. In that aspect, Xiong et al. [19] reported automated segmentation of endometrial lesions. Using sagittal T2WI of 154 patients with combined stage IA and IB endometrial cancer, they constructed a DL model and found that the detection rates of uterine and endometrial lesions were 98.7% and 87.93%, respectively. The accuracy, sensitivity, specificity, and AUC were 86.9%, 81.0%, 91.7%, and 0.89, respectively. Two radiologists conducted diagnostic examinations, and the accuracy, sensitivity, specificity, and AUC were 75.6%, 66.7%, 85.7%, and 0.81, respectively. There was no significant difference in the results between the DL model and radiologists. In this study, the DL model automatically segmented the region of interest (ROI) and generated a uterine cavity line, whereas in DL studies prior to this study, the ROI was manually set by radiologists. Automatic segmentation is a useful technology for further development of AI without human labor.

Other reports on the automated segmentation of endometrial lesions have been published. The larger the diameter of an endometrial lesion, the greater the likelihood of myometrial invasion or lymph node metastasis [20], which is a prognostic marker in preoperative evaluation, and surgical strategies can be considered. Although three-dimensional (3D) evaluation of endometrial cancer may improve diagnostic outcomes, it is labor-intensive and unfeasible for clinical evaluation of 3D images. Hodneland et al. [21] used preoperative MRI of 139 patients with endometrial cancer to construct a DL model for the automated segmentation of endometrial lesions in 3D images. Automated segmentation was comparable to that of the radiologists in this study. It is expected to be useful for the prediction of deep myometrial invasion and lymph node metastasis without being labor-intensive, in addition to a two-dimensional evaluation.

In addition to the development of DL models, radiomics, a medical imaging technique that aims to extract quantitative features from diagnostic images using computer algorithms to improve clinical decision-making, has also been improved [22]. Lefebvre et al. [23] developed multiparametric MRI-based radiomics models to predict deep myometrial invasion, lymphovascular space invasion, tumor grade, and FIGO staging. Using T2WI, diffusion-weighted imaging, and dynamic contrast-enhanced MRI of 157 patients with FIGO stage IA-IVB endometrial cancer, they extracted 3D images of MRI parameters and constructed a prediction model based on the extraction of important parameters. The AUCs for the prediction of deep myometrial invasion, lymphovascular space invasion, tumor grade, and FIGO staging were 0.81, 0.80, 0.74, and 0.84, respectively. The development of these prediction models is expected to aid preoperative decision-making in the future.

Although less common than reports on MRI using AI, there have also been studies evaluating DL models using PET/CT. Tang et al. [24] used PET/CT images of 155 patients with stage IA–IIA endometrial cancer to construct a DL model for the segmentation of endometrial lesions using six neural networks. The dice and precision values of the DL models were 0.546–0.743 and 0.419–0.966, respectively. In clinical practice, PET/CT is a useful tool for MRI to diagnose endometrial cancer, and a more accurate DL model is expected to be developed for PET/CT in the future.

5.4 Advances in AI Research for the Pathological Diagnosis of Endometrial Cancer

Research on artificial intelligence has made remarkable progress in the pathological diagnosis of endometrial cancer. The use of AI in pathological diagnosis is important because of the shortage of pathologists [25]. In addition, because molecular classification has been introduced in the FIGO2023 classification, the labor required for diagnosis can be much higher for pathologists, which may affect the accuracy of diagnosis. Therefore, AI technology is expected to be helpful in the pathological diagnosis of endometrial cancer.

Sun et al. [26] have reported the usefulness of AI-assisted histopathological diagnosis of endometrial cancer. In this study, a DL model was constructed using hematoxylin and eosin (H&E)-stained slides of endometrial cancer and non-cancerous tissues, such as normal endometrial tissue and atypical endometrial hyperplasia. The classification of the four categories—normal endometrium, endometrial polyps, endometrial hyperplasia, and endometrial adenocarcinoma— achieved an accuracy of 76.91%. In addition, the results of the classification of two categories, cancer and non-cancer, had an accuracy of 93.53%, an AUC of 0.9579, a sensitivity of 81.04%, and a specificity of 94.78%. Zhao et al. [27] constructed a DL model to identify precancerous lesions of endometrial intraepithelial neoplasia (EIN) in endometrial tissues. The diagnosis of EIN had an AUC of 0.9902. Furthermore, in the validation dataset of 135 specimens with 1631 slides, the accuracy of endometrial hyperplasia was 95.34%, which was equivalent to that of a mid-level pathologist (95.71%). In this study, a global-to-local multiscale convolutional neural network was used to construct the DL model, which combines the global and local feature extractor models. The global feature extractor extracts the cytological changes relative to the background, and the local feature extractor model extracts the gland-to-stromal ratio and lesion dimensions. The quality of the diagnosis was improved by evaluating both global and local features. Zhang et al. [28] also reported the use of AI for the diagnosis of H&E-stained whole-slide images (WSI). In this study, a DL model was constructed using H&E-stained WSI from 601 patients with endometrial cancer and non-cancerous tissue, including normal endometrial tissue and atypical endometrial hyperplasia. This DL model achieved a high diagnostic rate, with a sensitivity of 92.4%, a specificity of 80.1%, and an AUC of 0.928. Furthermore, it was performed on a prospective dataset with a sensitivity of 93.4%, a specificity of 83.7%, and an AUC of 0.933. In this DL model, segmentation of the

ROI with a heat map was possible, suggesting the possibility of supporting the diagnosis of pathologists. In fact, the DL model was able to identify even tiny lymphovascular invasions that could be missed by humans, and the high sensitivity of this model supports its usefulness as an AI-assisted diagnostic tool.

Recently, molecular classification into four categories according to prognosis using the TCGA database has been proposed: polymerase epsilon (*POLE*) (ultramutated), high microsatellite instability (MSI-H) (hypermutated) type, copy-number high type, and copy-number low type [29]. For the pathological diagnosis of endometrial cancer, FIGO2023 introduced a classification that takes molecular classification into four categories: *POLE* mutation type, mismatch repair deficient (MMRd) type, no specific molecular profile (NSMP) type, and *p53* mutation type (see Chap. 4). The introduction of immune checkpoint inhibitors is particularly beneficial for MSI-H and MMRd endometrial cancer, and clinical trials, such as the RUBY trial [30] and NRG-GY018 trial [31], have demonstrated that the MMRd population has a better prognosis than the mismatch repair deficient proficient (MMRp) population, which is important when considering treatment strategies for endometrial cancer (see Chap. 4). To diagnose MSI-H and MMRd, the MMR proteins MutL homologue 1 (*MLH1*), MutS homologue 2 (*MSH2*), MutS homologue 6 (*MSH6*), and post-meiotic segregation increased 2 (*PMS2*) were identified by immunohistochemical staining, polymerase chain reaction, or next-generation sequencing (NGS) [32]. Although this molecular classification presents an accurate prognosis, it is time-consuming and costly in clinical practice. Recently, a method to diagnose MSI-H and MMRd in H&E-stained specimens using DL models was reported. Zhang et al. [33] developed a DL model using H&E-stained WSI of endometrial cancer from the TCGA database, 47 patients with MSI-H, and 45 patients with microsatellite stable (MSS). The results showed an accuracy of 80% and a sensitivity of 86%. In this study, to diagnose MSI-H more efficiently using WSI, the models were fused into three categories: adipose and mucinous tissue, stromal and muscle tissue, and tumor epithelial tissue. The models were successfully integrated to improve the performance of MSS and MSI-H classifications. Furthermore, Wang et al. [34] developed a DL model using H&E-stained WSI of 529 endometrial cancer patients from the TCGA database. This DL model showed an accuracy of 94% and a sensitivity of 100% in endometrioid carcinoma G1 or G2 and an accuracy of 84% and a sensitivity of 94% in endometrioid carcinoma G3. It has been reported that the sensitivity and specificity of immunohistochemistry (IHC) testing are 94% and 88%, respectively [35], suggesting that the detection rate of MSI-H with H&E-stained WSI may be equivalent to that of IHC. The strength of this DL model is that the diagnostic speed of each WSI is only 1.03 seconds, which can resolve the problems of time-consuming and labor-intensive molecular classification, and is expected to be introduced in clinical practice in the future. Fremond et al. [36] also reported the construction of a DL model for molecular classification based on the ProMisE classification using H&E-stained WSI of 2028 endometrial cancer patients. The AUCs of *POLE* mutation, MMRd, NSMP, and p53 mutations were 0.849, 0.844, 0.833, and 0.928, respectively. In this study, morphological features (inflammatory cell count, stromal cell count, tumor cell count, tumor nuclei size, and tumor nuclei shape) were evaluated to analyze the characteristics of each molecular classification and extract suggestive features, such as a high density of lymphocytes in *POLE* mutation type and MMRd, and strong nuclear atypia in *p53* mutation type.

In the molecular classification based on the TCGA database, the copy-number high type had the poorest prognosis among the four categories [29]. *p53* mutation was evaluated as a surrogate for the copy-number high type in the ProMisE classification adopted by FIGO2023. Hong et al. developed a DL model to diagnose copy-number high types based on H&E-stained specimens. In this study, the DL model was constructed in InceptionResnet [37] using 456 patients and 496 slides from the TCGA database and Clinical Proteomic Tumor Analysis Consortium. The DL model was evaluated not only for histopathology but also for copy-number high type and mutations in 18 common genes associated with endometrial cancer. The AUC for the endometrial and serous carcinoma classification was 0.969, and that for the copy-number high type was 0.934. In addition, they reported a high diagnostic rate for *POLE* mutations with an AUC of 0.89. This DL model is expected to assist pathologists and clinicians in the diagnosis and consideration of treatment strategies by providing not only histopathological classification but also various genetic mutations.

In the development of AI using histopathological specimens, there have been several reports not only for diagnosis but also for prognosis prediction. Jiménez-Sánchez et al. [38] focused on the association between immune cell subtype and prognosis. In this study, immunofluorescence-stained specimens were used to stain CD68+ macrophages, CD8+ T cells, FOXP3+ regulatory T cells, and PD-L1/PD-1 proteins, and a DL model was developed. The DL model predicted the recurrence rate in 250 patients with endometrial cancer, including 32 recurrent cases of stage I–II (FIGO2009) and endometrioid carcinoma G1–G2. The results showed that the accuracy of recurrence prediction had an AUC of 0.90. In this study, the tumor microenvironment, including local phenotypes, cellular neighborhoods, and tissue areas, was evaluated and converted into heat maps for visualization, which can be recognized and interpreted by humans. Volinsky-Fremond et al. [39] developed a prognostic model called histopathology-based endometrial cancer-tailored outcome risk (HECTOR). They developed a DL model using 2072 patients with stage I–III endometrial cancer (FIGO2009) by integrating four molecular classifications based on the ProMisE classification and clinical stage information. The DL model was used to calculate the HECTOR risk scores for each patient. The predicted outcomes for 160 and 151 patients in the two-test datasets and the Concordance Index were 0.828 and 0.815, respectively. Further classification into low-, intermediate-, and high-risk groups resulted in 10-year distant recurrence-free probabilities of 97.0%, 77.7%, and 58.1%, respectively. This DL model classification is expected to provide prognostic information and assist in decision-making regarding postoperative chemotherapy and radiotherapy. In the field of histopathology, the development of AI has been remarkable in recent years and further progress is expected in the future.

5.5 AI in Hysteroscopy

Hysteroscopes are useful medical devices because they allow direct observation of uterine cancer lesions. For this reason, there have been several reports on the diagnosis of uterine cancer using AI-assisted hysteroscopes. Zhang et al. trained a VGGNet-16 model using 1851 still images from 454 patients to classify conditions such as endometrial hyperplasia, atypical endometrial hyperplasia, endometrial cancer, endometrial polyps, and uterine fibroids. The model achieved an average accuracy of 80.8%, surpassing the accuracy of three experienced physicians [40].

Takahashi et al. developed an AI model trained on videos to address the challenge of a limited number of cases. This study analyzed 177 cases performed using hysteroscopy at the University of Tokyo Hospital, including the normal endometrium (60 cases), uterine fibroids (21 cases), endometrial polyps (60 cases), atypical endometrial hyperplasia (15 cases), and endometrial cancer (21 cases). The videos of these 177 cases were converted into approximately 400,000 still images, which were then classified into malignant (atypical endometrial hyperplasia and endometrial cancer) and non-malignant (normal endometrium, endometrial polyps, and uterine fibroids) groups for deep learning analysis.

Three network models, Xception, MobileNetV2, and EfficientNetB0, were used for training, and their performances were compared. The initial evaluation using the standard algorithm achieved an accuracy rate of approximately 80%, with no significant differences observed between the network models. To improve diagnostic accuracy with a limited number of cases, a new algorithm called the "sequential method" was developed. This method classifies a case as malignant if 50 or more consecutive images are identified as malignant. Using this algorithm, the diagnostic accuracy was improved to over 85%. Additionally, a network model combination method was developed to enhance the accuracy further. Using this method, the diagnostic accuracy exceeded 90% [41].

5.6 Future Directions

AI research on endometrial cancer has been extensively conducted, primarily focusing on imaging and pathological diagnosis. However, it remains insufficient for clinical application, and several challenges must be addressed: (1) data scarcity and bias, (2) the need for clinical studies to validate its effectiveness, (3) the "black box" problem due to the lack of transparency in AI's diagnostic reasoning, (4) ensuring accuracy sufficient for clinical use, and (5) addressing ethical and legal concerns. We believe that AI-powered medical devices capable of practical clinical use can be developed for endometrial cancer by overcoming these challenges.

References

1. Russell S, Norvig P. Artificial intelligence: a modern approach. Englewood Cliffs: Prentice Hall; 1995.
2. Luger G, Stubblefield W. Artificial intelligence: structures and strategies for complex problem solving. San Francisco: Benjamin/Cummings; 2004.
3. Sone K, Toyohara Y, Taguchi A, et al. Application of artificial intelligence in gynecologic malignancies: a review. J Obstet Gynaecol Res. 2021;47(8):2577–85.
4. Hamamoto R, Suvarna K, Yamada M, et al. Application of artificial intelligence technology in oncology: towards the establishment of precision medicine. Cancers (Basel). 2020;12(12):3532.
5. Yamada M, Saito Y, Imaoka H, Saiko M, Yamada S, Kondo H, Takamaru H, Sakamoto T, Sese J, Kuchiba A, Shibata T, Hamamoto R. Development of a real-time endoscopic image diagnosis support system using deep learning technology in colonoscopy. Sci Rep. 2019;9(1):14465.
6. Fang S, Yang J, Wang M, et al. An improved image classification method for cervical precancerous lesions based on shufflenet. Comput Intell Neurosci. 2022;2022:9675628.
7. Chen X, Pu X, Chen Z, et al. Application of efficientNet-B0 and GRU-based deep learning on classifying the colposcopy diagnosis of precancerous cervical lesions. Cancer Med. 2023;12(7):8690–9.
8. Shinohara T, Murakami K, Matsumura N. Diagnosis assistance in colposcopy by segmenting acetowhite epithelium using u-net with images before and after acetic acid solution application. Diagnostics (Basel). 2023;13(9):1596.
9. Kim S, An H, Cho HW, et al. Pivotal clinical study to evaluate the efficacy and safety of assistive artificial intelligence-based software for cervical cancer diagnosis. J Clin Med. 2023;12(12):4024.
10. Jinnai S, Yamazaki N, Hirano Y, Sugawara Y, Ohe Y, Hamamoto R. The development of a skin cancer classification system for pigmented skin lesions using deep learning. Biomolecules. 2020;10:1123.
11. Bryant M, Massat MB. The ubiquity of AI at RSNA 2019. Appl Radiol. 2020;49:32–8.
12. Goldenthal IL, Sciacca RR, Riga T, Bakken S, Baumeister M, Biviano AB, Dizon JM, Wang D, Wang KC, Whang W, Hickey KT, Garan H. Recurrent atrial fibrillation/flutter detection after ablation or cardioversion using the AliveCor KardiaMobile device: iHEART results. J Cardiovasc Electrophysiol. 2019 Nov;30(11):2220–28.
13. Ben M'Barek IB, Jauvion G, Vitrou J, et al. DeepCTG® 1.0: an interpretable model to detect fetal hypoxia from cardiotocography data during labor and delivery. Front Pediatr. 2023;11:1190441.
14. Zhang Y, Gong C, Zheng L, Li X, Yang X. Deep learning for intelligent recognition and prediction of endometrial cancer. J Healthc Eng. 2021 Aug;26(2021):1148309.
15. Urushibara A, Saida T, Mori K, Ishiguro T, Inoue K, Masumoto T, Satoh T, Nakajima T. The efficacy of deep learning models in the diagnosis of endometrial cancer using MRI: a comparison with radiologists. BMC Med Imaging. 2022 Apr 30;22(1):80.
16. Fujii S, Matsusue E, Kigawa J, Sato S, Kanasaki Y, Nakanishi J, Sugihara S, Kaminou T, Terakawa N, Ogawa T. Diagnostic accuracy of the apparent diffusion coefficient in differentiating benign from malignant uterine endometrial cavity lesions: initial results. Eur Radiol. 2008 Feb;18(2):384-9.
17. Dong H, Dong H, Yu M, Lin Y, Chang C. Using deep learning with convolutional neural network approach to identify the invasion depth of endometrial cancer in myometrium using MR images: a pilot study. Int J Environ Res Public Health. 2020;17(16):5993.
18. Tao J, Wang Y, Liang Y, Zhang A. Evaluation and monitoring of endometrial cancer based on magnetic resonance imaging features of deep learning. Contrast Media Mol Imaging. 2022 Mar;18(2022):5198592.
19. Xiong L, Chen C, Lin Y, Mao W, Song Z. A computer-aided determining method for the myometrial infiltration depth of early endometrial cancer on MRI images. Biomed Eng Online. 2023 Oct 31;22(1):103.

20. Ytre-Hauge S, Husby JA, Magnussen IJ, Werner HM, Salvesen ØO, Bjørge L, Trovik J, Stefansson IM, Salvesen HB, Haldorsen IS. Preoperative tumor size at MRI predicts deep myometrial invasion, lymph node metastases, and patient outcome in endometrial carcinomas. Int J Gynecol Cancer. 2015 Mar;25(3):459–66.
21. Hodneland E, Dybvik JA, Wagner-Larsen KS, Šoltészov V, Munthe-Kaas AZ, Fasmer KE, Kraksta C, Lundervold A, Lundervold AS, Salvese Ø, Erickson BJ, Haldorsen I. Automated segmentation of endometrial cancer on MR images using deep learning. Sci Rep. 2021 Jan 8;11(1):179.
22. Shrestha P, Poudyal B, Yadollahi S, Wright DE, Gregory AV, Warner JD, Korfiatis P, Green IC, Rassier SL, Mariani A, Kim B, Laughlin-Tommaso SK, Kline TL. A systematic review on the use of artificial intelligence in gynecologic imaging – background, state of the art, and future directions. Gynecol Oncol. 2022 Sep;166(3):596–605.
23. Lefebvre TL, Ueno Y, Dohan A, Chatterjee A, Vallières M, Winter-Reinhold E, Saif S, Levesque IR, Zeng XZ, Forghani R, Seuntjens J, Soyer P, Savadjiev P, Reinhold C. Development and validation of multiparametric MRI-based radiomics models for preoperative risk stratification of endometrial cancer. Radiology. 2022 Nov;305(2):375–86. https://doi.org/10.1148/radiol.212873.
24. Tang D, Li C, Tianmin D, Jiang H, Ma D, Ma Z, Grzegorzek M, Jiang T, Sun H. ECPC-IDS: a benchmark endometrial cancer PET/CT image dataset for evaluation of semantic segmentation and detection of hypermetabolic regions. Comput Biol Med. 2024 Mar;171:108217.
25. Metter DM, Colgan TJ, Leung ST, Timmons CF, Park JY. Trends in the US and Canadian pathologist workforces from 2007 to 2017. JAMA Netw Open. 2019 May 3;2(5):e194337.
26. Sun H, Zeng X, Tao X, Peng G, Ma Y. Computer-aided diagnosis in histopathological images of the endometrium using a convolutional neural network and attention mechanisms. IEEE J Biomed Health Inform. 2020 Jun;24(6):1664–76.
27. Zhao F, Dong D, Hongyan D, Guo Y, Xue S, Wang Z, Xie X, Wang M, Zhang H, Cao X, He X. Diagnosis of endometrium hyperplasia and screening of endometrial intraepithelial neoplasia in histopathological images using a global-to-local multi-scale convolutional neural network. Comput Methods Prog Biomed. 2022 Jun;221:106906.
28. Zhang X, Ba W, Zhao X, Wang C, Li Q, Zhang Y, Lu S, Wang L, Wang S, Song Z, Shen D. Clinical-grade endometrial cancer detection system via whole-slide images using deep learning. Front Oncol. 2022 Nov 2:12:1040238.
29. Cancer Genome Atlas Research Network, Kandoth C, Schultz N, Cherniack AD, Akbani R, Liu Y, Shen H, Robertson AG, Pashtan I, Shen R, Benz CC, Yau C, Laird PW, Ding L, Zhang W, Mills GB, Kucherlapati R, Mardis ER, Levine DA. Integrated genomic characterization of endometrial carcinoma. Nature. 2013 May 2;497(7447):67–73.
30. Mirza MR, Chase DM, Slomovitz BM, dePont Christensen R, Novák Z, Black D, Gilbert L, Sharma S, Valabrega G, Landrum LM, Hanker LC, Stuckey A, Boere I, Gold MA, Auranen A, Pothuri B, Cibula D, McCourt C, Raspagliesi F, Shahin MS, Gill SE, Monk BJ, Buscema J, Herzog TJ, Copeland LJ, Tian M, He Z, Stevens S, Zografos E, Coleman RL, Powell MA; RUBY Investigators. Dostarlimab for Primary Advanced or Recurrent Endometrial Cancer. N Engl J Med. 2023;388(23):2145–58.
31. Eskander RN, Sill MW, Beffa L, Moore RG, Hope JM, Musa FB, Mannel R, Shahin MS, Cantuaria GH, Girda E, Mathews C, Kavecansky J, Leath CA 3rd, Gien LT, Hinchcliff EM, Lele SB, Landrum LM, Backes F, O'Cearbhaill RE, Al Baghdadi T, Hill EK, Thaker PH, John VS, Welch S, Fader AN, Powell MA, Aghajanian C. Pembrolizumab plus Chemotherapy in Advanced Endometrial Cancer. N Engl J Med. 2023 Jun 8;388(23):2159-2170.
32. Saeed OAM, Mann SA, Luchini C, Huang K, Zhang S, Sen JD, Piredda ML, Wang M, Baldrige LA, Matthew Sperling R, Curless KL, Cheng L. Evaluating mismatch repair deficiency for solid tumor immunotherapy eligibility: immunohistochemistry versus microsatellite molecular testing. Hum Pathol. 2021 Sep;115:10–8.
33. Zhang Y, Chen S, Wang Y, Li J, Kai X, Chen J, Zhao J. Deep learning-based methods for classification of microsatellite instability in endometrial cancer from HE-stained pathological images. J Cancer Res Clin Oncol. 2023 Sep;149(11):8877–88.

34. Wang C-W, HikamMuzakky NPF, Liu T-C, Lai P-J, Wang Y-C, Yu M-H, Chao T-K. Deep learning to assess microsatellite instability directly from histopathological whole slide images in endometrial cancer. NPJ Digit Med. 2024 May 29;7(1):143.
35. Stjepanovic N, Moreira L, Carneiro F, Balaguer F, Cervantes A, Balmaña J, Martinelli E, ESMO guidelines committee. Hereditary gastrointestinal cancers: ESMO clinical practice guidelines for diagnosis, treatment and follow-up. Ann Oncol. 2019 Oct 1;30(10):1558–71.
36. Fremond S, Andani S, Wolf JB, Dijkstra J, Melsbach S, Jobsen JJ, Brinkhuis M, Roothaan S, Jurgenliemk-Schulz I, Lutgens LCHW, Nout RA, van der Steen-Banasik EM, de Boer SM, Powell ME, Singh N, Mileshkin LR, Mackay HJ, Leary A, Nijman HW, Smit VTHBM, Creutzberg CL, Horeweg N, Koelzer VH, Bosse T. Interpretable deep learning model to predict the molecular classification of endometrial cancer from haematoxylin and eosin-stained whole-slide images: a combined analysis of the PORTEC randomised trials and clinical cohorts. Randomized Controlled Trial Lancet Digit Health. 2023 Feb;5(2):e71–82.
37. Hong R, Liu W, DeLair D, Razavian N, Fenyö D. Predicting endometrial cancer subtypes and molecular features from histopathology images using multi-resolution deep learning models. Cell Rep Med . 2021 Sep 23;2(9):100400.
38. Jiménez-Sánchez D, López-Janeiro Á, Villalba-Esparza M, Ariz M, Kadioglu E, Masetto I, Goubert V, Lozano MD, Melero I, Hardisson D, Ortiz-de-Solórzano C, de Andrea CE. Weakly supervised deep learning to predict recurrence in low-grade endometrial cancer from multiplexed immunofluorescence images. NPJ Digit Med. 2023 Mar 23;6(1):48.
39. Volinsky-Fremond S, Horeweg N, Andani S, Barkey Wolf J, Lafarge MW, de Kroon CD, Ørtoft G, Høgdall E, Dijkstra J, Jobsen JJ, Lutgens LCHW, Powell ME, Mileshkin LR, Mackay H, Leary A, Katsaros D, Nijman HW, de Boer SM, Nout RA, de Bruyn M, Church D, Smit VTHBM, Creutzberg CL, Koelzer VH, Bosse T. Prediction of recurrence risk in endometrial cancer with multimodal deep learning. Nat Med. 2024 Jul;30(7):1962-1973.
40. Zhang YZ, Wang ZH, Zhang J, Wang CC, Wang YS, Chen H, Shan LH, Hu JN, Gu JH, Ma X. Deep learning model for classifying endometrial lesions. J Transl Med. 2021 Jan 6;19(1):10.
41. Takahashi Y, Sone K, Noda K, Yoshida K, Toyohara Y, Kato K, Inoue F, Kukita A, Taguchi A, Nishida H, Miyamoto Y, Tanikawa M, Tsuruga T, Iriyama T, Nagasaka K, Matsumoto Y, Hirota Y, Hiraike-Wada O, Oda K, Maruyama M, Osuga Y, Fujii T. Automated system for diagnosing endometrial cancer by adopting deep-learning technology in hysteroscopy. PLoS One. 2021 Mar 31;16(3):e0248526.

Lymphadenectomy for Endometrial Cancer

6

Yosuke Konno

Abstract

Background
Endometrial cancer (EC) is commonly diagnosed at an early stage, allowing for curative surgical intervention. Standard management involves total hysterectomy, bilateral salpingo-oophorectomy, and staging based on the FIGO system, which emphasizes evaluation of pelvic and para-aortic lymph nodes to inform prognosis and adjuvant treatment planning.

Lymph Node Dissection: Diagnostic and Therapeutic Roles
Lymphadenectomy provides essential information for staging and risk assessment. While retrospective studies (e.g., SEPAL) support a therapeutic role for para-aortic dissection in high-risk patients, randomized trials such as ASTEC have not demonstrated a survival benefit for pelvic dissection, fueling ongoing debate.

Current Clinical Trials
JCOG1412 and ECLAT are investigating whether systematic lymphadenectomy improves survival, using rigorous design and quality control.

Tailored Approaches for Low- and High-Risk Patients
Omission of lymphadenectomy is feasible in low-risk cases using imaging and biomarkers. In contrast, systemic therapy may be prioritized in high-risk disease.

Future Perspectives
Technological advances such as sentinel node mapping and molecular diagnostics are expected to personalize surgical strategies.

Y. Konno (✉)
Department of Obstetrics and Gynecology, Hokkaido University Graduate School of Medicine and Faculty of Medicine, Sapporo, Japan
e-mail: konsuke1@med.hokudai.ac.jp

H. Watari, T. Mitamura (eds.), *Uterine Body Neoplasms*, Comprehensive Gynecology and Obstetrics, https://doi.org/10.1007/978-981-95-0946-1_6

Conclusion
Although survival benefit remains controversial, lymphadenectomy is integral to staging and risk stratification in EC. Emerging evidence and molecular tools will optimize its application within individualized treatment paradigms.

Keywords

Endometrial cancer · Clinical trial · Lymphadenectomy · Survival

6.1 Introduction

The treatment of endometrial cancer (EC) can be broadly classified into surgical therapy, radiotherapy, chemotherapy, and hormone therapy. In recent years, novel molecularly targeted agents—such as immune checkpoint inhibitors (ICIs) and tyrosine kinase inhibitors (TKIs)—have been introduced and applied to EC. However, as around 80% of cases are diagnosed early and are amenable to complete surgical resection, surgery remains the mainstay of initial treatment because of its reliability compared with the less predictable efficacy of radiotherapy or chemotherapy.

Accurate staging is pivotal in determining the optimal therapeutic strategy. Unlike cervical cancer, for which radiotherapy often serves as a primary treatment, the stage of EC is established postoperatively through pathological evaluation of the resected specimens. Therefore, the surgical procedures required for definitive staging are an intrinsic part of managing EC. In 1988, the International Federation of Gynecology and Obstetrics (FIGO) adopted a surgical staging system that requires evaluation of the uterus, bilateral adnexa, and the relevant retroperitoneal lymph nodes [1]. This comprehensive assessment, particularly of the para-aortic and pelvic lymph nodes, is indispensable for precise staging and ensures a more informed determination of subsequent therapeutic interventions. However, the absence of a clear description regarding the methodology of lymph node assessment led to some confusion in subsequent clinical practice. The most accurate way to determine whether lymph node metastasis (LNM) is present is to perform a systematic lymphadenectomy followed by pathological examination. At the opposite extreme is the approach of palpating or biopsying only the enlarged lymph nodes during surgery. In recent years, sentinel lymph node (SLN) navigation surgery has become increasingly popular in Western countries, and it is acknowledged as an acceptable option for early-stage cases in both the NCCN Guidelines [2] and the ESGO-ESTRO-ESP Guidelines [3]. In Japan, however, SLN navigation surgery has not yet been widely adopted in clinical practice, partly due to insurance coverage issues. Consequently, in actual clinical settings, the extent and thoroughness of lymph node dissection (LND) are carried out at varying levels among different institutions. In the new FIGO 2023 staging classification, molecular features have been incorporated; however, LNM remains categorized as stage IIIC, and the importance of diagnosing nodal involvement is unchanged [4]. Over the years, experts have debated which patients should have LND, how it should be performed, and how extensive it should be. In practice, there is still no clear consensus.

In discussing LND, it is essential to distinguish between its diagnostic role (staging, prognostication, and determining the need for adjuvant therapy) and its therapeutic role (whether removal of lymph nodes can prolong survival). Numerous reports have confirmed that LNM is a poor prognostic factor [5–7], thus establishing the diagnostic importance of lymph node assessment. However, regarding its therapeutic benefit, despite previous large-scale multicenter randomized controlled trials (RCTs), the presence of multiple unresolved issues has impeded the establishment of a definitive conclusion. Currently, two large prospective clinical trials are underway to evaluate the therapeutic significance of LND in patients with EC who are at risk of nodal metastasis: one in Japan (JCOG1412) [8] and another led by the German Arbeitsgemeinschaft für GynäkologischeOnkologie (AGO) group (ECLAT) [9]. Furthermore, if accurate preoperative assessment indicates a low risk of nodal involvement in early-stage disease, lymphadenectomy can be safely omitted. This approach not only minimizes complications but also shortens operative times and optimizes the use of healthcare resources.

This chapter provides an overview of the current status of LND in EC. Detailed information on SLN navigation surgery and minimally invasive laparoscopic or robotic procedures will be discussed in other sections.

6.2 Which Lymph Nodes Should Be Dissected? Regional Lymph Nodes in EC

When determining which lymph nodes to remove to evaluate metastases in EC, it is important to identify the regional lymph nodes (the first site of lymphatic drainage). Theoretically, a thorough dissection of these regional nodes leads to a more accurate diagnosis of LNM. This perspective is consistent with reports showing that, in EC, more lymph node metastases are detected when the number of dissected lymph nodes increases [10] and when the extent of dissection is broadened [11]. In the eighth edition of the UICC (Union for International Cancer Control) TNM classification [12], the regional lymph nodes for EC are defined as the pelvic and para-aortic lymph nodes. The area is specified as extending from the supra-inguinal lymph nodes (part of external iliac lymph nodes located superior to the inguinal ligament) on the peripheral side to the para-aortic lymph nodes below the renal vein on the central side (Fig. 6.1). Reports on the frequency of LNM in the pelvic and para-aortic regions vary widely. These variations can be attributed to differences in clinical and pathological backgrounds, surgical techniques for LND, and pathological examination methods. Some studies suggest that isolated para-aortic LNM is rare and may be omitted [13]. In a study of 281 intermediate- to high-risk patients who underwent systematic pelvic and para-aortic LND, para-aortic LNM was observed in about 15%. Among those with lymph node involvement, metastasis was confined to the pelvic nodes in 33%, involved both pelvic and para-aortic nodes in 51%, and was limited to the para-aortic nodes in 16%. Furthermore, approximately

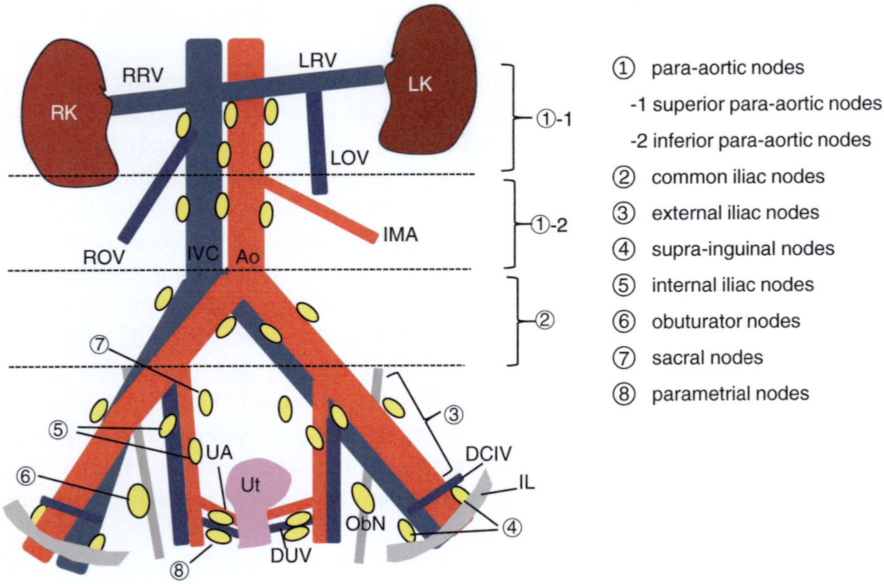

Fig. 6.1 Regional lymph nodes and their nomenclature in endometrial cancer. Ao aorta, DCIV deep circumflex iliac vein, DUV deep uterine vein, IL inguinal ligament, IMA inferior mesenteric artery, IVC inferior vena cava, LK left kidney, LOV left ovarian vein, LRV left renal vein, ObN obturator nerve, RK right kidney, ROV right ovarian vein, RRV right renal vein, UA uterine artery

80% of para-aortic LNM were located in the region cranial to the inferior mesenteric artery (IMA), showing a significantly higher frequency than in the para-aortic region below the IMA [14]. This study was conducted at a single institution (Mayo Clinic) with a limited number of surgeons and standardized surgical procedures. As a result of rigorous quality assurance, the frequency of metastasis detection was likely higher, and metastasis detection was more accurate. Even though the detection rate of para-aortic SLN varies depending on whether the tracer is injected into the cervix or the uterine corpus, it has been reported that para-aortic lymph nodes can serve as SLN in EC [15, 16]. These findings indicate that, despite their distance from the primary lesion, para-aortic lymph nodes remain regional lymph nodes for EC. They also confirm that the assessment of nodal metastases in the para-aortic region—especially above the IMA—cannot be overlooked for diagnostic purposes.

6.3 Omission of Lymphadenectomy in Low-Risk EC

Surgical cancer treatment must ensure curative outcomes while also minimizing invasiveness and postoperative complications. If the therapeutic effect is equivalent, a less invasive procedure is preferable. Performing a retroperitoneal LND prolongs operative time, increases blood loss, and raises the risk of complications such as postoperative ileus, thromboembolism, and prolonged hospitalization. It also elevates the likelihood of postoperative lower-limb lymphedema, lymphoceles, and lymphatic abscesses. SLN navigation surgery is an effective method to avoid unnecessary systematic LND. However, it can only be performed at a limited number of facilities. If we can preoperatively identify cases that do not require LND and thus omit a systematic dissection, it would offer significant benefits for both patients and healthcare economics.

In the GOG-33 study reported in 1987 [17], among cases clinically presumed to be confined to the uterine corpus, pelvic LNM was observed in 9% and para-aortic LNM in 6%. The study also showed that the risk of LNM could be classified using three postoperative pathological factors: histological grade, depth of myometrial invasion, and presence of intra-abdominal disease. In cases without intraperitoneal disease, 44 patients who had grade 1 tumors confined to the endometrium showed no LNM in either the pelvic or para-aortic regions. Among patients with grade 1 tumors and less than one-third of myometrial invasion, LNM was found in only 3% of pelvic nodes and 1% of para-aortic nodes. These results suggest that there is a subgroup for whom omitting LND may be reasonable. However, because this classification is based on postoperative pathology, it cannot be directly applied in actual clinical practice, where decisions about whether to perform LND are made preoperatively or intraoperatively. Preoperative factors reported to predict LNM include the histological grade and subtype based on biopsy, the extent of myometrial invasion on MRI, the tumor volume estimated by a volume index, and serum CA125 levels. Several scoring systems combining these factors have been proposed. At our institution, we use the LNM score proposed by Todo et al. [18], which incorporates histologic type/grade, serum CA125 level, and a volume index representing tumor volume. By combining this score with an MRI-based evaluation of myometrial invasion to assess the risk of LNM, we reported that LND could safely be omitted without significant problems [19]. Between 2003 and 2015, 110 patients at our institution did not undergo LND for various reasons. We evaluated the validity of three approaches [20]: our institution's LNM score combined with MRI-based assessment of myometrial invasion [19], the Korean KGOG low-risk criteria [21], and the KCC score proposed by the Kanagawa Cancer Center [22]. These scoring systems differ in several aspects, including the reference value for serum CA125, the cutoff for tumor volume, and the grading of endometrioid carcinoma. Using the LNM score, 64 patients met the criteria and there was 1 recurrence (1.6%). Under the KGOG low-risk criteria, 91 patients qualified—more cases than under the LNM score—but there were 3 recurrences (3.3%). Under the KCC score, only 40 patients qualified and none had a recurrence (Table 6.1). None of the cases that fulfilled all three scoring systems demonstrated lymph node recurrence, suggesting that these criteria for omitting LND are generally appropriate. Stricter criteria may reduce recurrence but also limit eligible cases, highlighting the need for better scoring systems. In the future, integrating genetic expression analysis of preoperative biopsy samples may improve risk assessment and allow selective omission of LND.

Table 6.1 Comparison of three preoperative scoring systems in our cohort

	LNM score + MI	KGOG low-risk criteria	KCC score
Preoperative factors			
Histological type/grade	Endometrioid G1/G2	Endometrioid any grade	Endometrioid G1
Serum CA125 (IU/ml)			
Premenopausal	<70	<35	<70
Postmenopausal	<28	<35	<25
MRI findings			
MI	No/minimal	<1/2	<1/2
Tumor volume	<36 cm^3	Not included	<6 cm^3
Matched patients	64 (58.2%)	91 (82.7%)	40 (36.4%)
Median follow-up period (month) [range]	60.5 [3–155]	60 [3–155]	63 [5–155]
Follow-up >3 years (*N*)	44 (69%)	65 (71%)	26 (65%)
Post-operative findings			
Histological type/grade			
Endometrioid G1/G2	62 (97%)	85 (93%)	38 (95%)
MI < ½	63 (98%)	90 (99%)	39 (98%)
LVSI (−)	63 (98%)	86 (95%)	39 (98%)
Pathological stage I	62 (97%)	88 (97%)	40 (100%)
Recurrence	1 (1.6%)	3 (3.3%)	0 (0%)
Recurrent site	Lung	Lung, peritoneum, ovary	–

MI myometrial invasion, *LVSI* lymphovascular space invasion, *LNM* lymph node metastasis

6.4 Therapeutic Significance of Lymphadenectomy in EC

The therapeutic value of systematic lymphadenectomy in EC, particularly in the pelvic and para-aortic regions, has been debated for many years. The key question is whether extensive lymph node removal improves survival and reduces recurrence. Removing a higher number of lymph nodes is useful for diagnosing LNM, as previously mentioned. Additionally, some studies suggest that removing more lymph nodes is associated with longer survival [10, 23–25]. The association between the number of resected lymph nodes and improved prognosis in node-negative cases may be due to the removal of undetected occult micrometastases, thereby reducing the risk of recurrence. Another potential explanation is the phenomenon known as "stage migration." Stage migration refers to the reclassification of cases from stage I or II to stage III when a greater number of lymph nodes are removed, allowing metastatic lymph nodes that might have been missed with fewer dissections to be identified. As a result, patients in the group with a higher number of dissected lymph nodes are more likely to represent true stage I or II cases. Conversely, the group with fewer dissected lymph nodes may include a certain proportion of stage III cases with poor prognosis, mistakenly classified as stage I or II (Fig. 6.2). This mixing of stages can lead to differences in survival rates between the groups. Consequently, this makes it difficult to attribute the observed survival benefits solely to the

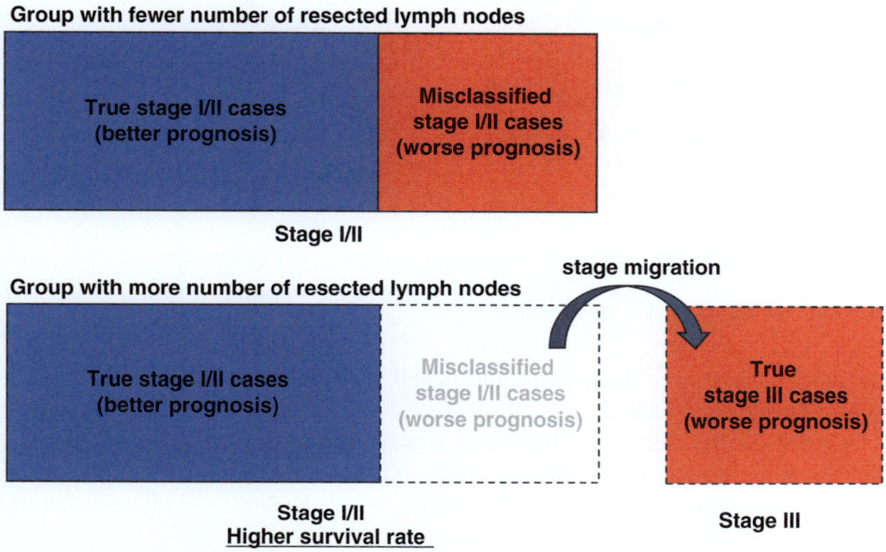

Fig. 6.2 Stage migration phenomenon in lymphadenectomy

therapeutic effect of lymphadenectomy. When evaluating the therapeutic effects of lymphadenectomy in EC, it is essential to consider both the risk of LNM and the impact of adjuvant therapies. In cases with a low risk of LNM, survival outcomes remain favorable even without lymphadenectomy, making it difficult to demonstrate its therapeutic benefit. On the other hand, in high-risk cases, adjuvant therapies are typically administered. Since the type of adjuvant therapy varies, its influence cannot be disregarded. Furthermore, lymphadenectomy is a form of localized treatment. In groups receiving radiotherapy—another localized treatment—as part of their adjuvant therapy, the therapeutic effect of lymphadenectomy may be masked. Consequently, studies focusing on patient groups treated with chemotherapy as their adjuvant therapy are more likely to provide clear evidence of the therapeutic benefits of lymphadenectomy. Recently, we reported on the therapeutic impact of pelvic lymphadenectomy (PLX) using the JGOG2043 cohort [26], which evaluated optimal chemotherapy regimens for high-risk EC patients following surgery [27]. Among 250 patients who underwent PLX as part of a unified chemotherapy regimen, despite varying adjuvant therapies, 83 patients had fewer than 20 pelvic lymph nodes (PLNs) removed, while 167 patients had 20 or more PLNs removed. There were no significant differences in patient demographics or LNM rates between the two groups. However, the group with 20 or more PLNs removed showed a trend toward reduced pelvic recurrence. Kaplan-Meier analysis revealed no statistically significant difference in survival rates between the two groups. Nevertheless, multivariate analysis identified the removal of 20 or more PLNs as an independent prognostic factor for overall survival (OS), comparable to advanced stage, high-risk histological subtypes, and advanced age (Table 6.2). However, the

Table 6.2 Multivariate analysis of prognostic factors for OS in endometrial cancer

	Hazard ratio	95% CI	*P* value
Number of resected PLN ≥20	0.49	0.24–0.99	0.048
Age ≥60	2.42	1.18–4.95	0.016
High-risk histology[a]	2.66	1.30–5.43	0.007
Stage III or IV	6.18	2.22–17.2	<0.001
Myometrial invasion ≥50%	2.16	0.95–6.95	0.067
LVSI positive	1.07	0.49–2.35	0.867

CI confidence interval, *LVSI* lymph vascular space invasion, *OS* overall survival, *PLN* pelvic lymph node

[a]Serous, clear cell, or grade 3 endometrioid

potential influence of "stage migration" cannot be entirely excluded, even for independent prognostic factors.

To evaluate the therapeutic significance of lymphadenectomy in EC while minimizing differences in patient backgrounds and eliminating unknown confounding factors, randomized controlled trials (RCTs) with standardized adjuvant therapy and consistent surgical quality are considered more reliable.

The ASTEC trial conducted in the UK sought to assess the therapeutic value of PLX in EC [28]. The study concluded that PLX did not improve prognosis, suggesting no therapeutic benefit. However, several limitations were noted in this trial. These included a low median number of dissected lymph nodes (12), the absence of para-aortic lymphadenectomy (PALX), and the inclusion of a substantial proportion (approximately 40%) of early-stage endometrial cancer with only superficial myometrial invasion, which generally carries a low risk of LNM. Consequently, the trial did not establish a consensus regarding the omission of PLX in intermediate- and high-risk groups. In contrast, the SEPAL study, conducted in Japan, demonstrated the therapeutic significance of PALX in intermediate- and high-risk EC patients [7]. However, this study had limitations as well. It was a retrospective study, the postoperative treatments were not standardized, and there was a notable difference in the number of pelvic lymph nodes dissected between the PLX group and the PLX and PALX group. This evidence highlights the need for high-quality RCTs to clarify the therapeutic role of lymphadenectomy in EC, particularly for intermediate- and high-risk groups.

In Japan, the Japan Clinical Oncology Group (JCOG) 1412 trial (a phase III randomized study evaluating the therapeutic significance of PALX in EC patients at risk of LNM) is in progress, with our group serving as the trial coordinating office. Meanwhile, the ECLAT (Endometrial Cancer Lymphadenectomy Trial) is being conducted [9]. This trial compares two groups of clinical stage I–II patients undergoing either systematic lymphadenectomy from the pelvic to para-aortic regions or no lymphadenectomy at all, either preoperatively or after hysterectomy. For comparison, we have summarized JCOG1412 and ECLAT in Table 6.3. In the JCOG1412 trial, the extent of lymphadenectomy is strictly defined by the study protocol, and intraoperative photographs of the surgical field are collected to ensure quality of LND. A central review of these photographs is performed based on specific criteria (Fig. 6.3), ensuring the critical quality of lymphadenectomy is maintained. Although

Table 6.3 JCOG1412 vs ECLAT

	JCOG1412	ECLAT
Phase, design	III, superior	III, superior
Primary endpoint	OS	OS
Treatment arm	PLX vs PLX + PALX	No LNX vs PLX + PALX
Age (years)	20–75	18–75
Preoperative stage	IB, II, IIIA, IIIB, IIIC1[a]	IA (high-risk histology), IB, II[c]
Adjuvant therapy	Chemotherapy[b]	Vaginal brachytherapy + chemotherapy[b]
Target accrual	560	640
Sites	47 (Japan only)	58 (Germany), 2 (Japan), 1 (Korea)
Status	Enrollment completed (Nov. 2022)	Recruiting (Nov. 2024)
Follow-up	5 years	5 years

OS overall survival, *PLX* pelvic lymphadenectomy, *PALX* para-aortic lymphadenectomy, *LNX* lymphadenectomy

[a]Single metastasis in pelvic nodes, [b]Paclitaxel + carboplatin, [c]Registration of secondary surgery after hysterectomy is also acceptable

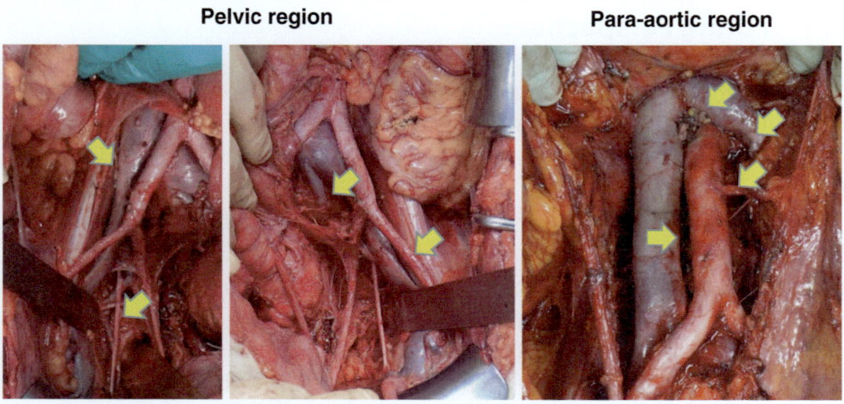

Fig. 6.3 Criteria for assessing the quality of lymphadenectomy. Pelvic region: (1) No residual tissues around iliac vessels and obturator nerve. (2) Easy identification of the surface of the sacrum. Para-aortic region: (1) Easy identification of the margin of the renal vein. (2) Easy identification of the stump of ovarian vessels. (3) No residual tissues around the inferior mesenteric artery. (4) Removal of inter-aortic nodes

interim results are still being analyzed, the median number of resected lymph nodes is reported to be 45 in the pelvic region and 26 in the para-aortic region, which is higher than previously reported. This suggests that the trial is adequately powered to evaluate the therapeutic significance of PALX. Similarly, in the ECLAT, the extent of lymphadenectomy is clearly illustrated in the protocol. Furthermore, the group conducting this trial previously managed the LION study [29], which evaluated the therapeutic role of lymphadenectomy in ovarian cancer. Given their experience, it is expected that the quality of lymphadenectomy in this trial meets or exceeds established standards.

In other cancers, retrospective studies have indicated the therapeutic significance of PALX. However, it has proven challenging to demonstrate its benefit in RCTs. For instance, in gastric cancer, the JCOG9501 trial was an RCT conducted with approximately 500 patients to evaluate the superiority of adding prophylactic PALX to standard D2 lymphadenectomy. However, no significant differences were observed in 5-year survival rates or recurrence patterns [30]. Similarly, in bladder cancer, an RCT involving about 400 patients with resectable muscle-invasive tumors compared limited lymphadenectomy (covering the obturator, internal, and external iliac regions) with extended lymphadenectomy (including the para-aortic nodes). The trial did not show significant differences in either disease-free survival (DFS) or OS [31].

However, it is worth noting that, unlike gastric and bladder cancer, the para-aortic lymph nodes are not considered SLN in these malignancies. This suggests the possibility of different outcomes in EC. The JCOG1412 trial, which completed patient enrollment in November 2022, aims to address this question. The results of this study are expected to be published after a 5-year follow-up period.

6.5 Future Perspectives

The increasing adoption of minimally invasive surgeries, such as laparoscopic procedures and SLN navigation surgery, is expected to make lymphadenectomy less invasive. However, the importance of tailoring lymphadenectomy to individual cases remains unchanged.

In our post hoc analysis of 384 cases from the JGOG2043 cohort, where PALX was performed, it was found that patients with two or more para-aortic LNM had significantly poorer survival outcomes compared to those with zero or one metastasis [32]. Notably, the prognosis for patients with one metastasis was similar to that of patients with no metastasis. This finding suggests that a combination of PALX and adjuvant chemotherapy could be effective in cases with only one metastasis. However, for cases with two or more para-aortic node metastases, the disease is likely systemic, and this treatment combination is less effective. This highlights the urgent need for novel therapeutic approaches.

For these high-risk patients with poor prognosis, several strategies are being considered. These include combining more intensive surgical interventions with chemotherapy and immune checkpoint inhibitors or prioritizing systemic therapies, such as antibody-drug conjugates, cytotoxic chemotherapy, or immune checkpoint inhibitors, instead of surgery. These approaches may offer the potential to improve outcomes in this difficult-to-treat population.

The JCOG1412 trial incorporates preoperative imaging data and clinicopathological factors, ensuring the accuracy of LNM diagnosis by maintaining an adequate number of lymph nodes resected. Using this high-quality cohort, genetic expression analyses of tumor tissues are planned to identify biomarkers associated with LNM and prognosis.

In the future, gene expression analyses of biopsy samples, combined with preoperative imaging and clinicopathological factors, are expected to enable more accurate predictions of individual LNM risk. If we can also predict whether lymphadenectomy will improve prognosis in specific cases, it will pave the way for truly individualized lymphadenectomy strategies.

References

1. Shepherd JH. Revised FIGO staging for gynaecological cancer. Br J ObstetGynaecol. 1989;96:889–92.
2. NCCN clinical practice guidelines in oncology (NCCN guidelines)—[uterine neoplasms]. Version 1.2025. Available at: https://www.nccn.org/professionals/physician_gls/pdf/uterine.pdf (Accessed [Jan 12, 2025]).
3. Concin N, Matias-Guiu X, Vergote I, Cibula D, Mirza MR, Marnitz S, Ledermann J, Bosse T, Chargari C, Fagotti A, Fotopoulou C, Gonzalez Martin A, Lax S, Lorusso D, Marth C, Morice P, Nout RA, O'Donnell D, Querleu D, Raspollini MR, Sehouli J, Sturdza A, Taylor A, Westermann A, Wimberger P, Colombo N, Planchamp F, Creutzberg CL. ESGO/ESTRO/ESP guidelines for the management of patients with endometrial carcinoma. Int J Gynecol Cancer. 2021 Jan;31(1):12–39.
4. Berek JS, Matias-Guiu X, Creutzberg C, Fotopoulou C, Gaffney D, Kehoe S, et al. FIGO staging of EC: 2023. J Gynecol Oncol. 2023;34:e85.
5. Garg G, Morris RT, Solomon L, Toy EP, Kruger M, Clary K, Munkarah A, Wright JD. Evaluating the significance of location of LNM and extranodal disease in women with stage IIIC EC. Gynecol Oncol. 2011 Nov;123(2):208–13.
6. Hiura M, Nogawa T, Matsumoto T, Yokoyama T, Shiroyama Y, Wroblewski J. Long-term survival in patients with Para-aortic lymph node metastasis with systematic retroperitoneal lymphadenectomy followed by adjuvant chemotherapy in endometrial carcinoma. Int J Gynecol Cancer. 2010 Aug;20(6):1000–5.
7. Todo Y, Kato H, Kaneuchi M, Watari H, Takeda M, Sakuragi N. Survival effect of para-aortic lymphadenectomy in endometrial cancer (SEPAL study): a retrospective cohort analysis. Lancet. 2010 Apr 3;375(9721):1165–72.
8. Watari H, Katayama H, Shibata T, Ushijima K, Satoh T, Onda T, Aoki D, Fukuda H, Yaegashi N, Sakuragi N, Gynecologic cancer study Group of the Japan Clinical Oncology Group. Phase III trial to confirm the superiority of pelvic and para-aortic lymphadenectomy to pelvic lymphadenectomy alone for endometrial cancer: Japan clinical oncology group study 1412 (SEPAL-P3). Jpn J Clin Oncol. 2017 Oct 1;47(10):986–90.
9. Emons G, Kim JW, Weide K, de Gregorio N, Wimberger P, Trillsch F, Gabriel B, Denschlag D, Kommoss S, Aydogdu M, Papathemelis T, Gropp-Meier M, Muallem MZ, Kühn C, Müller A, Frank M, Weigel M, Bronger H, Lampe B, Rau J, Schade-Brittinger C, Harter P. EC Lymphadenectomy Trial (ECLAT) (pelvic and PALX in patients with stage I or II EC with high risk of recurrence; AGO-OP.6). Int J Gynecol Cancer. 2021 Jul;31(7):1075–9.
10. Chan JK, Urban R, Cheung MK, Shin JY, Husain A, Teng NN, et al. Lymphadenectomy in endometrioid uterine cancer staging: how many lymph nodes are enough? A study of 11,443 patients. Cancer. 2007;109:2454–60.
11. Huang M, Chadha M, Musa F, Friedmann P, Kolev V, Holcomb K. Lymph nodes: is total number or station number a better predictor of LNM in EC? Gynecol Oncol. 2010 Nov;119(2):295–8.
12. Sobin LH, Gospodarowicz MK, Wittekind C, editors. TNM atlas: illustrated guide to the TNM classification of malignant tumours. 7th ed. Wiley-Blackwell: Chichester; 2015.

13. Baiocchi G, Faloppa CC, Mantoan H, Camarço WR, Badiglian-Filho L, Kumagai LY, De Brot L, da Costa AABA. PALX can be omitted in most EC patients at risk of LNM. J Surg Oncol. 2017 Aug;116(2):220–6.
14. Mariani A, Dowdy SC, Cliby WA, Gostout BS, Jones MB, Wilson TO, Podratz KC. Prospective assessment of lymphatic dissemination in EC: a paradigm shift in surgical staging. Gynecol Oncol. 2008 Apr;109(1):11–8.
15. Niikura H, Okamura C, Utsunomiya H, Yoshinaga K, Akahira J, Ito K, Yaegashi N. Sentinel lymph node detection in patients with EC. Gynecol Oncol. 2004 Feb;92(2):669–74.
16. Rossi EC, Kowalski LD, Scalici J, Cantrell L, Schuler K, Hanna RK, Method M, Ade M, Ivanova A, Boggess JF. A comparison of sentinel lymph node biopsy to lymphadenectomy for EC staging (FIRES trial): a multicentre, prospective, cohort study. Lancet Oncol. 2017 Mar;18(3):384–92.
17. Creasman WT, Morrow CP, Bundy BN, Homesley HD, Graham JE, Heller PB. Surgical pathologic spread patterns of EC. A Gynecologic Oncology Group Study. Cancer. 1987 Oct 15;60(8 Suppl):2035–41.
18. Todo Y, Okamoto K, Hayashi M, Minobe S, Nomura E, Hareyama H, Takeda M, Ebina Y, Watari H, Sakuragi N. A validation study of a scoring system to estimate the risk of lymph node metastasis for patients with endometrial cancer for tailoring the indication of lymphadenectomy. Gynecol Oncol. 2007 Mar;104(3):623–8.
19. Mitamura T, Watari H, Todo Y, Kato T, Konno Y, Hosaka M, et al. Lymphadenectomy can be omitted for low-risk EC based on preoperative assessments. J Gynecol Oncol. 2014;25:301–5.
20. Konno Y, Asano H, Shikama A, Aoki D, Tanikawa M, Oki A, Horie K, Mitsuhashi A, Kikuchi A, Tokunaga H, Terao Y, Satoh T, Ushijima K, Ishikawa M, Yaegashi N, Watari H. Lymphadenectomy issues in EC. J Gynecol Oncol. 2021 Mar;32(2):e25.
21. Kang S, Nam JH, Bae DS, Kim JW, Kim MH, Chen X, No JH, Lee JM, Kim JH, Watari H, Kim SM, Kim SH, Seong SJ, Kim KT, Kim SC, Kim JH, Lim MC, Lee JY, Ryu SY, Yang B, Kim BG. Preoperative assessment of LNM in EC: a Korean Gynecologic oncology group study. Cancer. 2017 Jan 1;123(2):263–72.
22. Imai K, Kato H, Katayama K, Nakanishi K, Kawano A, Iura A, Konnai K, Onose R, Hirahara F, Miyagi E. A preoperative risk-scoring system to predict LNM in EC and stratify patients for lymphadenectomy. Gynecol Oncol. 2016 Aug;142(2):273–7.
23. Abu-Rustum NR, Iasonos A, Zhou Q, Oke E, Soslow RA, Alektiar KM, Chi DS, Barakat RR. Is there a therapeutic impact to regional lymphadenectomy in the surgical treatment of endometrial carcinoma? Am J Obstet Gynecol. 2008 Apr;198(4):457.e1–5.
24. Seagle BL, Gilchrist-Scott D, Graves S, Strohl AE, Nieves-Neira W, Shahabi S. Association of lymph node count and overall survival in node-negative ECs. JCO Clin Cancer Informat. 2017;1 https://doi.org/10.1200/CCI.16.00064.
25. Lutman CV, Havrilesky LJ, Cragun JM, Secord AA, Calingaert B, Berchuck A, et al. Pelvic lymph node count is an important prognostic variable for FIGO stage I and II endometrial carcinoma with high-risk histology. Gynecol Oncol. 2006;102:92–7.
26. Nomura H, Aoki D, Michimae H, Mizuno M, Nakai H, Arai M, et al. Effect of taxane plus platinum regimens vs doxorubicin plus cisplatin as adjuvant chemotherapy for EC at a high risk of progression: a randomized clinical trial. JAMA Oncol. 2019;5:833–40.
27. Konno Y, Mayama M, Takehara K, Yokoyama Y, Suzuki J, Susumu N, Harano K, Nakagawa S, Nakanishi T, Yamagami W, Yoshihara K, Nomura H, Okamoto A, Aoki D, Watari H. Prognostic impact of the number of resected pelvic nodes in EC: Japanese Gynecologic oncology group study JGOG2043 post hoc analysis. J Gynecol Oncol. 2024; https://doi.org/10.3802/jgo.2025.36.e3. Epub ahead of print

28. ASTEC Study Group, Kitchener H, Swart AM, Qian Q, Amos C, Parmar MK. Efficacy of systematic PLX in EC (MRC ASTEC trial): a randomized study. Lancet. 2009;373:125–36.
29. Harter P, Sehouli J, Lorusso D, Reuss A, Vergote I, Marth C, et al. A randomized trial of lymphadenectomy in patients with advanced ovarian neoplasms. N Engl J Med. 2019;380:822–32.
30. Sasako M, Sano T, Yamamoto S, Kurokawa Y, Nashimoto A, Kurita A, Hiratsuka M, Tsujinaka T, Kinoshita T, Arai K, Yamamura Y, Okajima K, Japan Clinical Oncology Group. D2 lymphadenectomy alone or with para-aortic nodal dissection for gastric cancer. N Engl J Med. 2008 Jul 31;359(5):453–62.
31. Gschwend JE, Heck MM, Lehmann J, Rübben H, Albers P, Wolff JM, Frohneberg D, de Geeter P, Heidenreich A, Kälble T, Stöckle M, Schnöller T, Stenzl A, Müller M, Truss M, Roth S, Liehr UB, Leißner J, Bregenzer T, Retz M. Extended versus limited LND in bladder cancer patients undergoing radical cystectomy: survival results from a prospective, randomized trial. Eur Urol. 2019 Apr;75(4):604–11.
32. Konno Y, Mayama M, Takehara K, Yokoyama Y, Suzuki J, Susumu N, Harano K, Nakagawa S, Nakanishi T, Yamagami W, Yoshihara K, Nomura H, Okamoto A, Aoki D, Watari H. Prognostic significance of Para-aortic node metastasis in EC: Japanese Gynecologic oncology group study JGOG2043 post hoc analysis. J Gynecol Oncol. 2025;36:e57.

Sentinel Node Navigation Surgery for Endometrial Cancer

7

Hideaki Yahata

Abstract

Endometrial cancer is the most common gynecologic malignancy. Primary surgery, including hysterectomy, bilateral salpingo-oophorectomy, and retroperitoneal lymph node dissection, is the standard procedure for endometrial cancer. The status of lymph node metastasis is important in determining clinical stage and the suitability for postoperative therapy. Pelvic and para-aortic lymph node dissection is necessary for accurate diagnosis of lymph node metastases. Some patients truly have lymph node metastases, but most patients undergo unnecessary lymph node dissection, which results in postoperative lymphedema and lymphangitis of the lower extremities. The sentinel lymph node (SLN) is the first lymph node to receive flow from the primary tumor. An SLN biopsy without systematic lymph node dissection can reduce surgical morbidities such as lymphedema, lymphocyst, and lymphangitis. It has been reported that SLN biopsies have a high detection rate, sensitivity, and negative predictive value in patients with endometrial cancer. Several studies suggest that an SLN biopsy is an effective treatment strategy and there is no significant difference between an SLN biopsy and lymphadenectomy in oncologic outcomes for the patients with endometrial cancer. Prospective randomized trials involving SLN biopsies in patients with endometrial cancer are ongoing, and promising results are expected.

Keywords

Endometrial cancer · Sentinel lymph node · Sentinel node navigation surgery

H. Yahata (✉)
Department of Obstetrics and Gynecology, Graduate School of Medical Sciences, Kyushu University, Fukuoka, Japan
e-mail: yahata.hideaki.134@m.kyushu-u.ac.jp

H. Watari, T. Mitamura (eds.), *Uterine Body Neoplasms*, Comprehensive Gynecology and Obstetrics, https://doi.org/10.1007/978-981-95-0946-1_7

7.1 Introduction

Endometrial cancer is the most common gynecologic malignancy in Japan, and according to the 2021 Japan Society of Obstetrics and Gynecology Annual Patient Report, 13,912 patients were treated for endometrial cancer [1]. The Global Cancer Observatory reported in 2020 that 417,367 patients had newly diagnosed endometrial cancer worldwide with 97,370 deaths [2]. Hysterectomy plus bilateral salpingo-oophorectomy with retroperitoneal lymph node dissection or biopsy is the standard treatment for endometrial cancer. Lymph node metastasis is an important risk factor and is essential for determining clinical stage and suitability for postoperative adjuvant treatment. However, lymph node dissection can cause postoperative lymphedema and lymphangitis of the lower extremities, which may reduce the quality of life. The lymph node metastasis rate for grade 1–2 endometrial cancer with <50% myometrial invasion and a tumor diameter <2 cm is only 1–2% [3, 4]. Indeed, patients with endometrial cancer who are node-negative do not benefit from unnecessary lymph node dissection.

The sentinel lymph node (SLN) is the first lymph node to receive flow from the primary tumor. If there is no evidence of metastases to the SLN, it is thought that there are no metastases to other lymph nodes. Therefore, if the SLN is confirmed to be metastasis-negative intraoperatively, sentinel node navigation surgery (SNNS) can be performed without lymph node dissection. SNNS is a highly beneficial procedure that shortens the operative time, decreases blood loss, and reduces postoperative lymphedema. The usefulness of SNNS has been reported, prospective clinical trials are ongoing, and promising results are expected. Omission of lymph node dissection as a clinical trial in facilities that are proficient in the technique of SLN biopsy and have the cooperation of pathologists for intraoperative diagnosis (proposal of implementation: level 2) has been proposed by the Japan Society of Gynecologic Oncology according to the treatment guidelines for endometrial cancer published in 2023. However, SLN biopsy for endometrial cancer is not currently covered by national public health insurance in Japan.

7.2 Types of Tracers

The tracers used to identify SLNs include the following: (1) a dye method using patent blue; (2) a radioisotope (RI) method using ^{99m}technetium-labeled phytate; and (3) a fluorescence method using indocyanine green (ICG).

The dye method uses patent blue or isosulfan blue, and when injected into a tumor, the SLN can be identified as a blue-colored lymph node. The dye method requires only a dye as a tracer, making the method easy and inexpensive to perform at any facility [5]. Because of the rapid flow of the dye, the dye quickly enters the lymph node, then drains out. Therefore, the dye must be injected while the lymph node that may correspond to the SLN is confirmed under direct line of sight of the injection; otherwise, the primary node may be missed or a secondary node may be misidentified as the SLN.

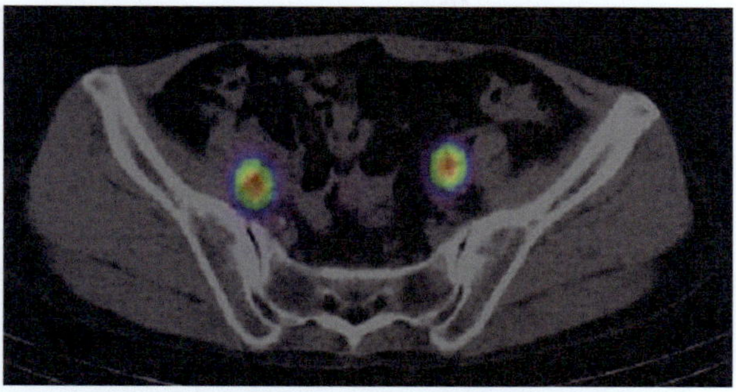

Fig. 7.1 Sentinel lymph node depicted by SPECT-CT

^{99m}Technetium-labeled phytate (200–1000 nm diameter) is used for the RI method [6]. The location of the SLN is estimated by lymphoscintigraphy and SPECT-CT, then the SLN is identified intraoperatively with gamma probes. SPECT-CT facilitates intraoperative exploration because the three-dimensional location of the SLN site is easy to find (Fig. 7.1). In addition, because the particle size is large, the particles remain in the corresponding lymph node for a long time. Unlike the dye method, it is possible to search for SLNs using the RI method over a long period of time. Of note, the requirement for RI equipment limits the number of facilities that can perform this technique.

The ICG fluorescence method generally involves injection of ICG, and the SLN can be identified as bright nodes using a fluorescent camera [7]. The ICG fluorescence method is more sensitive than the dye method, allows visualization of lymphatic vessels beneath the tissue surface, and enables identification of SLNs over time. However, unlike breast cancer and malignant melanoma, all lymphatic channels from the primary tumor to the pelvic lymph nodes cannot be visualized. In recent years, the ICG fluorescence method has been increasingly used in laparoscopic and robot-assisted surgery. The SLN is easy to identify with DaVinci Xi using the Firefly mode (Fig. 7.2).

Currently, only ^{99m}technetium-labeled phytate is covered by insurance in Japan. There is an urgent need for insurance coverage of the ICG fluorescence method.

7.3 Site of Injection

Administration of tracers has been reported at the following sites: hysteroscopic submucosal administration in the uterine cavity [8]; intraoperative administration in the submucosa of the uterine serosa [6]; and transvaginal administration in the uterine cervix [9].

Tracer administration to the uterine cervix is the most common method among the administration methods described above (Fig. 7.3). The National Comprehensive Cancer Network (NCCN) guidelines suggest that the tracer should be administered to the cervix. Although the advantages of this method are its simplicity and reproducibility, administration of tracer is not considered appropriate for detection of the para-aortic lymph node region and should be considered a method of administration

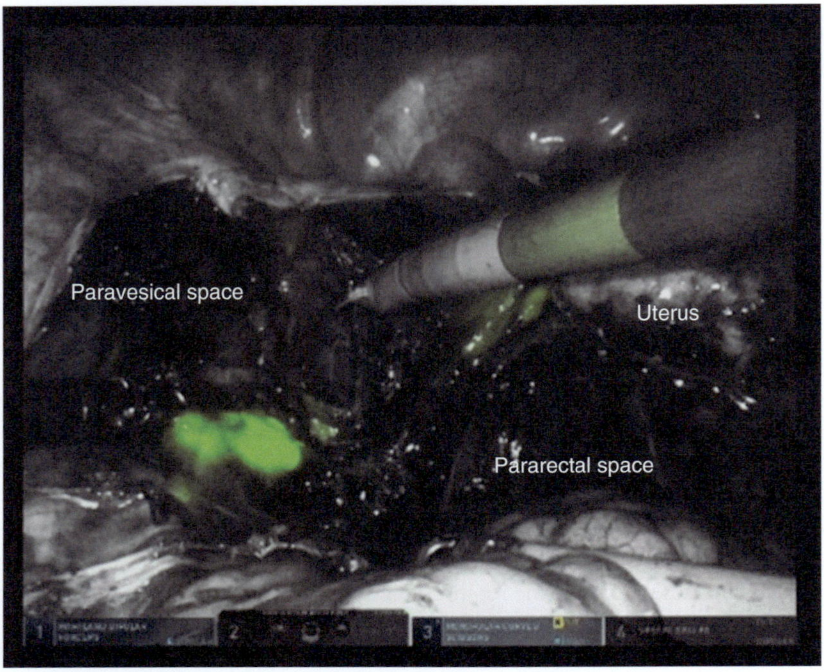

Fig. 7.2 Fluorescent sentinel lymph node in Firefly mode

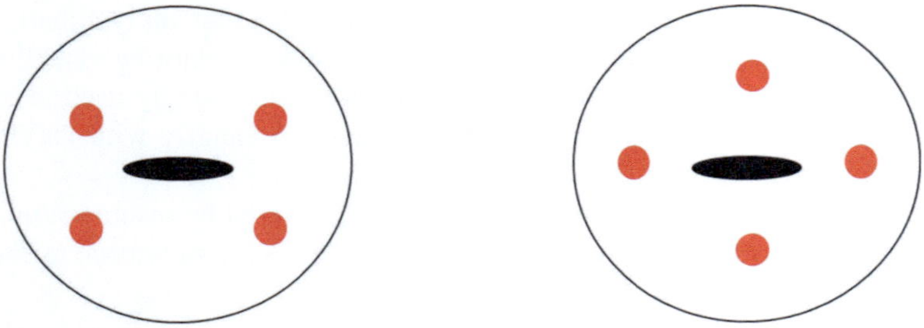

Fig. 7.3 Cervical injection sites (2, 4, 8, and 10 o'clock or 0, 3, 6, and 9 o'clock)

limited to detection of SLNs in the pelvis. A combination of injections into the superficial (1–3 mm) and deep (1–2 cm) layers of the uterine cervix has been reported to deliver tracers to the main layers of the cervical and body lymphatic pathway origins. Lymphatic flow from the uterine body crosses the umbilical artery, to the external iliac lymph nodes, internal iliac lymph nodes, and obturator lymph nodes. The lymphatic flow also courses along the mesoureter to the common iliac lymph nodes and sacral lymph nodes [10]. A meta-analysis involving 55 studies also examined differences in detection rates by method of administration. Tracer administration to the uterine cervix significantly increased the detection rate of SLNs in the pelvic region and bilateral detection rate compared to administration to the uterine body, while the SLN detection rate in the para-aortic region was significantly decreased [11].

Another site of administration is needed to identify the SLN in the para-aortic region. Kataoka et al. [8] reported that SLNs are present in the pelvic region and 37.1% of SLNs are in para-aortic lymph nodes; the sensitivity and negative predictive value were 100% each using hysteroscopic sub-endometrial injection with a RI and sub-serosal ICG injection. Niikura et al. [9] reported that the SLN detection rate was highest (96%) following cervical RI injection, but no SLNs were detected in the para-aortic area. Para-aortic SLNs were only detected by hysteroscopic RI injection (56%). A randomized controlled trial conducted by the MITO group showed that detection of SLNs in the para-aortic area was slightly higher in patients receiving a hysteroscopic injection, but the difference with the cervical route was not statistically significant [12].

The administration of tracer via the uterine cavity or uterine serosa may be considered in high-grade histologic subtype or deep myometrial invasion cases with a risk of metastasis to para-aortic lymph nodes.

7.4 Histologic Subtype

An SLN biopsy is commonly performed for low-grade histologic subtypes (grade 1–2 endometrioid carcinoma) confined to the uterus. The ESMO Clinical Practice Guideline shows that an SLN biopsy can be considered a strategy for nodal assessment in low- and intermediate-risk endometrioid carcinoma (e.g., stage IA grade 1–3 and stage IB grade 1–2) [13]. However, an SLN biopsy of the high-grade histologic subtype is also acceptable. A meta-analysis involving 429 cases in 9 studies of the stage I high-grade histologic subtype (grade 3 endometrioid carcinoma, serous carcinoma, clear cell carcinoma, and carcinosarcoma) showed that SLN identification has a sensitivity of 92%, false-negative rate of 8%, and negative predictive value of 97%, which indicated that an SLN biopsy is possible in the high-grade histologic subtype [14]. Altin et al. [15] reported that overall sensitivity, negative predictive value, and false-negative rate of an SLN biopsy were 81.8%, 95%, and 18.2%, respectively, and the sensitivity and negative predictive value improved to 96.4% and 98.9%, respectively, by using SLN algorithm steps in 244 patients with non-endometrioid high-grade carcinoma, or grade 3 endometrioid carcinoma. The NCCN guidelines have shown that an SLN biopsy is an acceptable alternative to systematic lymph node dissection for lymph node staging in high intermediate- and high-risk stage I–II endometrioid carcinoma based on the above reports [10].

7.5 Diagnosis of SLN Metastases

SLN metastases are classified according to metastasis size into macrometastases (MACs) >2 mm in size, micrometastases (MICs) between 0.2 and 2 mm in size, and isolated tumor cells (ITCs) <0.2 mm in size.

An intraoperative pathologic diagnosis of an SLN is usually performed by sectioning at 2 mm in the short axis and hematoxylin and eosin staining. This procedure makes SNNS possible if the SLN is negative for metastasis. However, MICs or ITCs might be determined by postoperative immunohistochemical staining. Ultra-staging is important to determine the correct FIGO stage and selection of appropriate postoperative adjuvant therapy. There is no standard protocol for ultra-staging for endometrial cancer. Chiang et al. [16] reported a protocol in which SLNs are cut into two adjacent paraffin block sections at two levels by 50 microns with two slides at each level stained with hematoxylin and eosin and cytokeratin AE1:AE3 immunohistochemistry. This protocol showed complete concordance in 91.8% of SLNs and a false-negative rate of 0–0.9% in detecting MICs and MACs.

It is controversial if ITCs should be identified using ultra-staging. A meta-analysis involving 286 patients in 8 studies (MICs, 187; ITCs, 99) reported a relative risk of recurrence of 1.34 (MICs, 1.07; ITCs, 1.67), which is a higher recurrence rate than lymph node-negative cases [17]. In contrast, patients with ITCs had no significant difference in survival compared to patients with negative lymph nodes. Nevertheless, the presence of ITCs had a significant association with receipt of adjuvant therapy, lymphovascular space invasion, and deep myometrial invasion ($P < 0.01$) in 1089 node-negative patients and 125 with ITCs [18]. ITCs are not reflected in the 2023 FIGO staging; therefore, the oncologic significance warrants further examination.

A new metastasis diagnostic method using one-step nucleic acid amplification (OSNA) has recently been attempted. The OSNA method solubilizes specimens and performs gene amplification, detection, and result determination in a single step and can be applied to cytokeratin 19 (CK19) mRNA detection in excised lymph nodes to diagnose lymph node metastases. Togami et al. [19] reported that the sensitivity and specificity of the OSNA assay were 0.918 (95% CI, 0.819–0.973) and 0.989 (95% CI, 0.973–0.997), respectively, in 437 lymph nodes from 133 patients with cervical and endometrial cancers. The OSNA assay may be useful for intraoperative diagnosis of lymph node metastases and is comparable to a histopathologic examination [19]. A large, observational, multicenter study (ENDO-OSNA) showed that the sensitivity of the OSNA assay was 92%, the specificity was 82%, the diagnostic accuracy was 83%, and the negative predictive value was 99% by OSNA for CK19 mRNA in 379 SLNs of 147 endometrial cancer patients [20].

7.6 SLN and FIGO 2023 Staging

The 2023 FIGO classification of stage IIIC endometrial cancer with lymph node metastasis was revised as follows [21]: IIIC, metastasis to the pelvic or para-aortic lymph nodes or both; IIIC1, metastasis to the pelvic lymph nodes; IIIC1i, MICs; IIIC1ii, MACs; IIIC2, metastasis to the para-aortic lymph nodes up to the renal veins with or without metastasis to the pelvic lymph nodes; IIIC2i, MICs; IIIC2ii, MACs; and ITCs are not considered in the 2023 FIGO staging system, but if ITCs are present in the SLN, the stage is pN0(i+).

7.7 Detection Rate, Sensitivity, and Negative Predictive Value

Several reports have been published about the detection rate, sensitivity, and negative predictive value. Detection rate means the proportion of patients in whom any SLN was identified with successful SLN mapping. Sensitivity means the percentage with true lymph node metastasis and correctly identified as SLN metastasis. Negative predictive value means the percentage of patients with a negative SLN who had no lymph node metastasis.

A prospective multicenter study (SENTI-ENDO trial) showed that the detection rate per patient was 89%, the sensitivity was 84%, and the negative predictive value was 97% in 125 patients with stage I–II endometrial cancer who underwent an SLN biopsy followed by pelvic lymph node dissection [22]. Barlin et al. [23] reported that the detection rate for finding at least one SLN was 81% (401/498 cases), the sensitivity was 85.1%, the false-negative rate was 14.9%, and the negative predictive value was 98.1% by using cervical injection of blue dye. The FIRES trial was a multicenter, prospective, cohort study patients with clinical stage I endometrial cancer of all histologic subtypes. The FIRES trial showed the detection rate for finding at least one SLN was 86% (293/385 cases), the sensitivity to detect node-positive disease was 97·2% (95% CI, 85.0–100), and the negative predictive value was 99·6% (95% CI, 97.9–100) [24]. The SHREC-trial for FIGO stage I–II high-risk endometrial cancer had a sensitivity of 98% (95% CI, 89–100) and a negative predictive value of 99.5% (95% CI, 97–100) by using cervical injection of ICG [25]. The SENTOR study, a prospective, multicenter cohort study for stage I grade 2 endometrioid or high-grade endometrial cancer reported a detection rate of 97.4% per patient (95% CI, 93.6–99.3%), a sensitivity of 96% (95% CI, 81–100%), and a negative predictive value of 99% (95% CI, 96–100%) [26].

7.8 Oncologic Outcome and SNNS

In a meta-analysis involving 5820 cases (SLN biopsy, 2152 cases; lymph node dissection, 3668 cases) in 15 studies, SLN biopsies detected significantly more positive pelvic lymph nodes than systematic lymph node dissection (OR, 1.91; 95% CI, 1.36–2.68; $P < 0.001$) and significantly reduced intraoperative bleeding (mean differences, −54.40; 95% CI, −23.45 to −85.36; $P < 0.001$). Moreover, there was no difference in overall survival (OS) ($I^2 = 79\%$, $P = 0.81$) and progression-free survival (PFS) ($I^2 = 52\%$, $P = 0.31$) between the two groups [27]. Six studies involving 3536 patients (SNNS, 1249 cases; lymph node dissection, 2287 cases) also found no difference between the two groups with respect to the overall recurrence rate (4.3% vs. 7.3%; OR, 0.90; 95% CI, 0.58–1.38; $P = 0.63$) or lymph node recurrence (1.2% vs. 1.7%; OR, 1.51; 95% CI, 0.70–3.29; $P = 0.29$) [28]. A national cancer database study showed that of 68,614 patients (SNNS, 1777 cases; lymph node dissection, 64,796 cases, both procedures, 2041 cases) with stage I–III endometrial cancer who underwent hysterectomy and lymph node assessment, neither SLN biopsy alone nor SLN biopsy followed by lymphadenectomy was associated with significantly different OS compared to lymphadenectomy alone (HR, 0.92; 95% CI, 0.73–1.17 and HR, 0.91; 95% CI, 0.77–1.08, respectively) based on multivariable analysis [29].

7.9 Ongoing Clinical Trials

Several prospective randomized trials with a focus on endometrial cancer are ongoing in which peritoneal lymph node dissection versus SLN biopsy only are compared.

In an Australian and New Zealand Gynaecological Oncology Group (ANZGOG) phase III randomized controlled trial (ENDO-3), the primary hypothesis for stage I is that SLN biopsy does not adversely affect patient outcomes (lymphedema, morbidity, and quality of life) and will not increase treatment-related morbidity or health service costs compared to patients without retroperitoneal node dissection after 12 months of surgery. The primary hypothesis for stage II is that patients without retroperitoneal lymph node dissection have a non-inferior disease-free survival (DFS) rate at 4.5 years after surgery compared to those who undergo SLN biopsy. Eligibility criteria include FIGO clinical stage I patients who undergo laparoscopic or robotic total hysterectomy and bilateral salpingo-oophorectomy with SLN biopsy or without retroperitoneal node dissection. In stage I, 444 patients will be enrolled to demonstrate feasibility and quality of life. If these criteria are met, stage II will enroll a further 316 patients [30].

In the Korean Gynecologic Oncology Group (KGOG) 2029/SELYE trial, the primary endpoint is 3-year DFS, and the secondary endpoints are 3-year OS, 5-year DFS, 5-year OS after surgery, pattern of recurrence, immediate surgical outcomes, success rate of SLN mapping, postoperative lymph-related complications, postoperative quality of life, and postoperative cost effectiveness. The inclusion criteria are FIGO clinical stage I–II endometrial cancer patients who plan to undergo laparoscopic or robotic hysterectomy and pelvic lymph node dissection, including all epithelial histologic sub-types. Patients with pelvic and para-aortic lymph nodes <15 mm in size on preoperative CT and MRI scans were included. A total of 810 patients (405 in each group) will be enrolled [31].

In the ALICE trial, the primary endpoint is to prove that SLN biopsy is oncologically non-inferior to systematic lymph node dissection. The secondary endpoints are 5-year OS, morbidity related to surgical procedures and early and late surgical complications, incidence of lower limb lymphedema, and quality of life related to surgical procedures. The inclusion criterion is patients without extrauterine disease, including all histologic sub-types or grade 1 or 2 endometrioid carcinoma with greater than one-half myometrial invasion or cervical invasion on imaging. A total of 356 cases (178 in each group) will be enrolled [32].

In the SNEC trial, the primary endpoint is the 2-year PFS. The secondary endpoints are the 5-year PFS, 5-year OS, adverse effects caused by surgery, and quality of life. The inclusion criterion is patients with endometrial cancer confined to the uterus with intermediate-to-high risk factors, including all histologic subtypes. Patients with grade 1–2 endometrioid carcinoma and pre-surgical endometrial lesions <2 cm in size with less than one-half myometrial invasion will be excluded. A total of 780 cases (390 in each group) will be enrolled from six hospitals in China within a 3-year period and followed up for 5 years [33].

In the INSEC trial, the primary endpoint is 3-year DFS. The secondary endpoint is OS and quality of life. Stage I endometrioid carcinoma patients are eligible. A total of 722 cases (361 in each group) will be enrolled from three hospitals in China [34].

7.10 Future and Prospects

SLN biopsy without retroperitoneal lymph node dissection is an efficacious procedure to prevent lymphedema and lymphangitis, resulting in an improved quality of life for the patient. Robotic and laparoscopic surgeries are becoming the mainstream in early-stage endometrial cancer, and SNNS is a compatible option in terms of minimally invasive surgery. Some prospective randomized studies previously mentioned will be expected to show promising data, i.e., no difference in oncologic outcome between SNNS and systematic lymph node dissection. SLN biopsy for endometrial cancer should be covered by national health insurance in Japan as soon as possible.

References

1. Kajiyama H, Tamauchi S, Takahashi F, et al. Annual report of the committee on gynecologic oncology, the Japan Society of Obstetrics and Gynecology: annual patient report for 2021 and annual treatment report for 2016. J Obstet Gynaecol Res. 2024; https://doi.org/10.1111/jog.16168.
2. Sung H, Ferlay J, Siegel RL, et al. Global cancer statistics 2020: GLOBOCAN estimates of incidence and mortality worldwide for 36 cancers in 185 countries. CA Cancer J Clin. 2021;71(3):209–49.
3. Mariani A, Dowdy SC, Cliby WA, et al. Prospective assessment of lymphatic dissemination in endometrial cancer: a paradigm shift in surgical staging. Gynecol Oncol. 2008;109(1):11–8.
4. Chi DS, Barakat RR, Palayekar MJ, et al. The incidence of pelvic lymph node metastasis by FIGO staging for patients with adequately surgically staged endometrial adenocarcinoma of endometrioid histology. Int J Gynecol Cancer. 2008;18(2):269–73.
5. Altgassen C, Pagenstecher J, Hornung D, et al. A new approach to label sentinel nodes in endometrial cancer. Gynecol Oncol. 2007;105(2):457–61.
6. Niikura H, Okamura C, Utsunomiya H, et al. Sentinel lymph node detection in patients with endometrial cancer. Gynecol Oncol. 2004;92(2):669–74.
7. Ruscito I, Gasparri ML, Braicu EI, et al. Sentinel node mapping in cervical and endometrial cancer: Indocyanine green versus other conventional dyes-a meta-analysis. Ann Surg Oncol. 2016;23(11):3749–56.
8. Kataoka F, Susumu N, Yamagami W, et al. The importance of para-aortic lymph nodes in sentinel lymph node mapping for endometrial cancer by using hysteroscopic radio-isotope tracer injection combined with subserosal dye injection: prospective study. Gynecol Oncol. 2016;140(3):400–4.

9. Niikura H, Kaiho-Sakuma M, Tokunaga H, et al. Tracer injection sites and combinations for sentinel lymph node detection in patients with endometrial cancer. Gynecol Oncol. 2013;131(2):299–303.
10. Uterine neoplasm (Version: 3.2024) NCCN clinical practice-guidelines in oncology. https://www.nccn.org/guidelines/category_1.
11. Bodurtha Smith AJ, Fader AN, Tanner EJ. Sentinel lymph node assessment in endometrial cancer: a systematic review and meta-analysis. Am J Obstet Gynecol. 2017;216(5):459–476.e10.
12. Ditto A, Casarin I, Pinelli C, et al. Hysteroscopic versus cervical injection for sentinel node detection in endometrial cancer: a multicenter prospective randomised controlled trial from the Multicenter Italian trials in ovarian cancer (MITO) study group. Eur J Cancer. 2020;140:1–10.
13. Oaknin A, Bosse TJ, Creutzberg CL, et al. Endometrial cancer: ESMO clinical practice guideline for diagnosis, treatment and follow-up. Ann Oncol. 2022;33(9):860–77.
14. Marchocki Z, Cusimano MC, Clarfield L, et al. Sentinel lymph node biopsy in high-grade endometrial cancer: a systematic review and meta-analysis of performance characteristics. Am J Obstet Gynecol. 2021;225(4):367.e1–367.e39.
15. Altın D, Taşkın S, Ortaç F, et al. Diagnostic accuracy of sentinel node biopsy in non-endometrioid, high-grade and/or deep myoinvasive endometrial cancer: a Turkish gynecologic oncology group study (TRSGO-SLN-006). Gynecol Oncol. 2022;164(3):492–7.
16. Chiang S, Tessier-Cloutier B, Klein E, et al. Establishing guidelines for sentinel lymph node ultrastaging in endometrial cancer. Int J Gynecol Cancer. 2024;34(5):681–8.
17. Gómez-Hidalgo NR, Ramirez PT, Ngo B, et al. Oncologic impact of micrometastases or isolated tumor cells in sentinel lymph nodes of patients with endometrial cancer: a meta-analysis. Clin Transl Oncol. 2020;22:1272–9.
18. Rios-Doria E, Abu-Rustum NR, Alektiar KM, et al. Prognosis of isolated tumor cells and use of molecular classification in early stage endometrioid endometrial cancer. Int J Gynecol Cancer. 2024;34(9):1373–81.
19. Togami S, Tanimoto A, Yanazume S, et al. Evaluation of the one-step nucleic acid amplification assay for detecting lymph node metastasis in patients with cervical and endometrial cancer: a multicenter prospective study. Gynecol Oncol. 2023;170:70–6.
20. Diestro MD, Berjón A, Zapardiel I, et al. One-step nucleic acid amplification (OSNA) of sentinel lymph node in early-stage endometrial cancer: Spanish Multicenter study (ENDO-OSNA). Cancers (Basel). 2021;13(17):4465.
21. Berek JS, Matias-Guiu X, Creutzberg C, et al. FIGO staging of endometrial cancer: 2023. Int J Gynaecol Obstet. 2023;162(2):383–94.
22. Ballester M, Dubernard G, Lécuru F, et al. Detection rate and diagnostic accuracy of sentinel-node biopsy in early stage endometrial cancer: a prospective multicentre study (SENTI-ENDO). Lancet Oncol. 2011;12(5):469–76.
23. Barlin JN, Khoury-Collado F, Kim CH, et al. The importance of applying a sentinel lymph node mapping algorithm in endometrial cancer staging: beyond removal of blue nodes. Gynecol Oncol. 2012;125(3):531–5.
24. Rossi EC, Kowalski LD, Scalici J, et al. A comparison of sentinel lymph node biopsy to lymphadenectomy for endometrial cancer staging (FIRES trial): a multicentre, prospective, cohort study. Lancet Oncol. 2017;18(3):384–92.
25. Persson J, Salehi S, Bollino M, et al. Pelvic sentinel lymph node detection in high-risk endometrial cancer (SHREC-trial)-the final step towards a paradigm shift in surgical staging. Eur J Cancer. 2019;116:77–85.
26. Cusimano MC, Vicus D, Pulman K, et al. Assessment of sentinel lymph node biopsy vs lymphadenectomy for intermediate- and high-grade endometrial cancer staging. JAMA Surg. 2021;156(2):157–64.
27. Gu Y, Cheng H, Zong L, et al. Operative and oncological outcomes comparing sentinel node mapping and systematic lymphadenectomy in endometrial cancer staging: meta-analysis with trial sequential analysis. Front Oncol. 2021;10:580128.
28. Bogani G, Murgia F, Ditto A, et al. Sentinel node mapping vs. lymphadenectomy in endometrial cancer: a systematic review and meta-analysis. Gynecol Oncol. 2019;153(3):676–83.

29. Garzon S, Mariani A, Day CN, et al. Overall survival after surgical staging by lymph node dissection versus sentinel lymph node biopsy in endometrial cancer: a national cancer database study. Int J Gynecol Cancer. 2022;32(1):28–40.
30. Obermair A, Nicklin J, Gebski V, et al. A phase III randomized clinical trial comparing sentinel node biopsy with no retroperitoneal node dissection in apparent early-stage endometrial cancer — ENDO-3: ANZGOG trial 1911/2020. Int J Gynecol Cancer. 2021;31(12):1595–601.
31. Park JY, Kim JH, Baek MH, et al. Randomized comparison between sentinel lymph node mapping using indocyanine green plus a fluorescent camera versus lymph node dissection in clinical stage I-II endometrial cancer: a Korean Gynecologic oncology group trial (KGOG2029/SELYE). J Gynecol Oncol. 2022;33(6):e73.
32. Baiocchi G, Andrade CEMC, Ribeiro R, et al. Sentinel lymph node mapping versus sentinel lymph node mapping with systematic lymphadenectomy in endometrial cancer: an open-label, non-inferiority, randomized trial (ALICE trial). Int J Gynecol Cancer. 2022;32(5):676–9.
33. Guan J, Xue Y, Zang RY, et al. Sentinel lymph node mapping versus systematic pelvic lymphadenectomy on the prognosis for patients with intermediate-high-risk endometrial cancer confined to the uterus before surgery: trial protocol for a non-inferiority randomized controlled trial (SNEC trial). J Gynecol Oncol. 2021;32(4):e60.
34. Guo Y, Sun L, Chen X, et al. A multicenter noninferior randomized controlled study of sentinel lymph node biopsy alone versus sentinel lymph node biopsy plus lymphadenectomy for patients with stage I endometrial cancer, INSEC trial concept. BMC Cancer. 2023;23(1):1184.

8 Minimally Invasive Surgery for Endometrial Cancer

Yoshito Terai

Abstract

In recent years, minimally invasive surgeries such as laparoscopic surgery and robot-assisted surgery have become more common than open surgery for uterine endometrial cancer. Fewer postoperative moderate or severe adverse events, shorter hospital stays, less pain, earlier resumption of normal activities, and improved quality of life with this minimally invasive surgery have been reported. In Japan, laparoscopic surgery for early-stage uterine endometrial cancer was introduced as an advanced medical treatment in August 2008; laparoscopic surgery was approved as an insurance-covered treatment for cases suspected of being stage IA uterine endometrial cancer, up to and including pelvic lymph node dissection in 2014; and laparoscopic para-aortic lymph node dissection for endometrial cancer became covered by health insurance in 2020. On the other hand, robot-assisted surgery for early-stage uterine endometrial cancer became covered by health insurance in 2018 and currently has become more common in Japan. In this chapter, we demonstrate the position of minimally invasive surgery at the present time and discuss its limitations in cases of endometrial cancer.

Keywords

Minimally invasive surgery · Laparoscopic surgery · Robot-assisted surgery · Para-aortic lymph node dissection · Obese patients · Hinotori™

Y. Terai (✉)
Department of Obstetrics and Gynecology, Kobe University Graduate School of Medicine, Kobe, Japan
e-mail: yterai@med.kobe-u.ac.jp

H. Watari, T. Mitamura (eds.), *Uterine Body Neoplasms*, Comprehensive Gynecology and Obstetrics, https://doi.org/10.1007/978-981-95-0946-1_8

8.1 Introduction

Endometrial cancer is among the most common gynecologic malignancies, and the number of cases has been increasing in recent years. In Japan, approximately 18,000 cases occur annually, and nearly 2600 people die from the disease. However, according to the 2020 Japan Society of Obstetrics and Gynecology Cancer Registry, 72.7% of cases are stage I, which is limited to the uterine corpus [1]. The main treatment for endometrial cancer is surgical intervention, and the standard procedure is a total hysterectomy, bilateral salpingo-oophorectomy, pelvic and para-aortic lymph node dissection (biopsy), and intraperitoneal cytology. In cases of early-stage cancer with a low risk of recurrence, a lymphadenectomy is performed up to the pelvic lymph nodes. Meanwhile, in cases of early-stage endometrial cancer with an intermediate to high risk of recurrence, such as special types of cancer (endometroid carcinoma (grade 3), serous carcinoma, clear cell carcinoma, or carcinosarcoma), cases with more than half muscle invasion, or cases with presumed vaginal involvement that had apparent uterine-confined disease, a lymphadenectomy is performed up to the para-aortic lymph nodes. When endometrial cancer patients receive open abdominal surgery, the surgical incision extends from the navel area to the subxiphoid area during a lymphadenectomy of the para-aortic lymph nodes, reaching a length of about 50 cm to the lower part of the xiphoid process (Fig. 8.1a). The larger the surgical incision with open surgery, the greater the surgical invasion, and therefore, the surgery requires caution against postoperative complications such as intraoperative

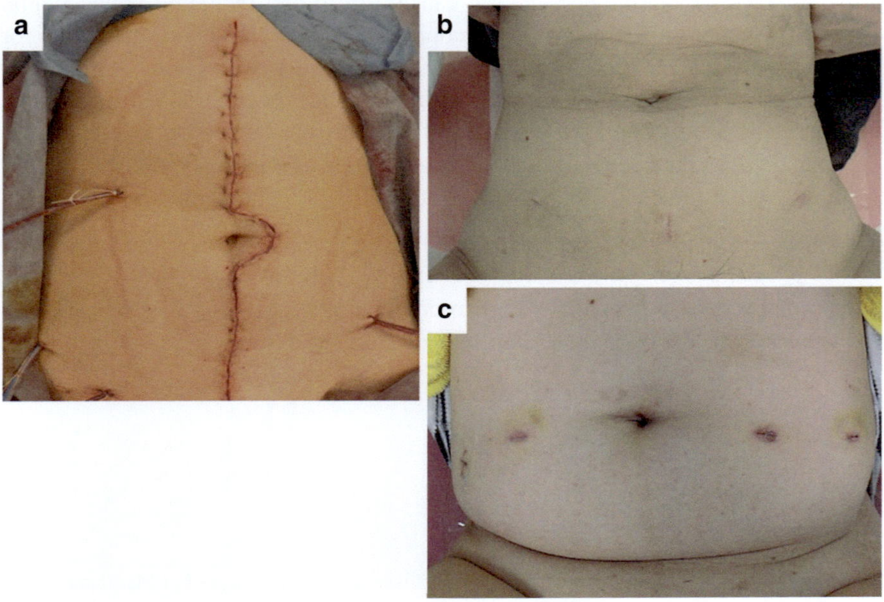

Fig. 8.1 Surgical incision for (**a**) open surgery, (**b**) laparoscopic surgery, and (**c**) robot-assisted surgery

bleeding and postoperative intestinal obstruction due to adhesion. On the other hand, laparoscopic surgery (Fig. 8.1b) and robot-assisted surgery (Fig. 8.1c) have become more common in recent years, and fewer postoperative moderate or severe adverse events, shorter hospital stays, less pain, earlier resumption of normal activities, and improved quality of life with this minimally invasive surgery have been reported [2]. In this chapter, we demonstrate the position of minimally invasive surgery at the present time and discuss its limitations in cases of endometrial cancer.

8.2 Laparoscopic Surgery for Uterine Corpus Cancer

Laparoscopic surgery for uterine corpus cancer was first reported by Childers et al. in 1992 [3]. It has become more widespread in Europe and the United States than open surgery, as it offers cosmetic benefits, reduced postoperative pain, and reduced intraoperative blood loss and blood transfusion rates. It also offers precise surgical manipulation with an enlarged field of vision and early postoperative recovery, along with improved QOL due to shortened hospital stays. In Japan, laparoscopic surgery for early-stage uterine endometrial cancer was introduced as an advanced medical treatment in August 2008, and laparoscopic surgery was approved as an insurance-covered treatment for cases suspected of being stage IA uterine endometrial cancer, up to and including pelvic lymph node dissection in 2014. Laparoscopic para-aortic lymph node dissection for intermediate- and high-risk stage I and II uterine cancer began in July 2017 under advanced medical treatment category A, and in 2020, laparoscopic para-aortic lymph node dissection for endometrial cancer became covered by health insurance (Fig. 8.2). In randomized controlled trials (RCTs) comparing laparoscopic surgery and open surgery for early-stage uterine endometrial cancer, laparoscopic surgery tended to be associated with longer operating times than open surgery in all of the studies [2, 4–6]. On the other hand,

Fig. 8.2 View of a para-aortic lymphadenectomy during laparoscopic surgery

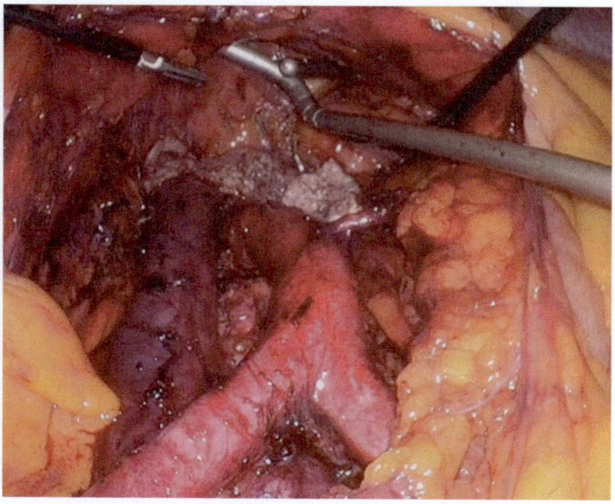

intraoperative blood loss tended to be significantly lower in laparoscopic surgery. In a systematic review of perioperative complications [7], it was found that the incidence of postoperative complications was significantly lower in laparoscopic surgery than in open surgery, but it should be noted that intraoperative complications, such as bladder-ureteral injury, intestinal injury, and vascular injury, were more common in laparoscopic surgery than in open surgery. The number of pelvic lymph nodes removed were 11.5–23.5 for laparoscopic surgery and 10.7–22.2 for open surgery, which were equivalent, and it is thought that laparoscopic pelvic lymph node dissection can be performed with the same curability as open surgery. The pelvic recurrence rate was also 1.3–5.0%, which was equivalent to the 1.0–15.8% rate for open surgery. In 2012, the prognostic analysis of the large-scale LAP2 study was published [8]. The average observation period was 59.3 (38–62.9) months, and the 3-year recurrence rate was 11.4%, which was not significantly different from the 10.2% for open surgery. The estimated 5-year survival rate was 89.8%, which was almost an identical rate for open surgery with 89.8%. In addition, the results of a retrospective study of 6304 patients aged 65 years or older with stage I to III uterine endometrial cancer from the US SEER database suggest that laparoscopic surgery and robot-assisted surgery have the same prognosis and curative effect as open surgery for early-stage uterine endometrial cancer [9]. In a cohort study in Japan, the operating time for laparoscopic surgery tended to be slightly longer than that for open surgery, but the intraoperative blood loss of 42.9 ± 76.3 ml was significantly less than the 236 ± 183 ml for open surgery ($p < 0.01$). In addition, the length of hospital stay was significantly shorter for laparoscopic surgery (9.3 ± 2 days) than for open surgery (14.4 ± 12.0 days) ($p < 0.01$). The duration of postoperative analgesic use was 1.4 ± 1 day for laparoscopic surgery and 2.4 ± 1.8 days for open surgery, with the duration being significantly shorter for laparoscopic surgery ($p < 0.01$), thus proving that laparoscopic surgery is minimally invasive. There was no significant difference between the two groups in the number of lymph nodes removed, with 32.3 ± 13.1 nodes removed in the laparoscopic surgery group and 28.0 ± 11.9 nodes removed in the open surgery group. The survival rate was 95.1% for disease-free survival and 97.3% for overall survival, with an observation period of around 2 years, and there was no difference from open surgery, as stated in other reports [10]. In the case of early-stage endometrial cancer with an intermediate to high risk of recurrence, the group that underwent a para-aortic lymph node dissection showed a significantly longer recurrence-free period and survival period than the group that underwent surgery up to pelvic lymph node dissection, due to the risk of lymph node metastasis, and it is thought that there is a therapeutic significance to para-aortic lymph node dissection [11]. In a multicenter retrospective study of intermediate- to high-risk early-stage endometrial cancer in Japan [12], the median intraoperative blood loss was 143 ml in the laparoscopic group (54 cases), while it was 988 ml in the open surgery group (99 cases), which was significantly lower in the laparoscopic group. There was no difference in intraoperative complications between the two groups, and postoperative complications of intestinal obstruction were significantly lower in the laparoscopic group, with no cases occurring in that group. The rate of lymphedema was lower in the laparoscopic group than in the open surgery group (3.7% versus 13.1%; $p = 0.07$), but the rate of chyle or lymphorrhea was higher in the laparoscopic group than in the open surgery group (18.5% versus 7.1%; $p = 0.02$). Similarly, the length of hospital stay was significantly shorter in the laparoscopic group. There was no significant difference in the operative time between the two groups, but the amount of intraoperative bleeding was lower in the laparoscopic group ($p < 0.01$). The number of para-aortic lymph nodes removed was lower in the laparoscopic group (26 ± 11) than in the open surgery group (31 ± 13) ($p = 0.02$), but the recurrence rate was 7.4% in the laparoscopic surgery group and 14% in the open surgery group. Although it was short term, there was no significant difference between the two groups. Moreover, in the 271 cases of laparoscopic para-aortic lymph node dissection performed under advanced medical care from 2017 to 2020 in Japan, intraoperative blood loss was 150 ± 250 ml, and the median number of para-aortic lymph nodes removed was 29 ± 15. There were also no serious complications, such as cases that required a conversion to open surgery. Intraoperative complications were observed in 4.8% of cases, including ureteral injury in 5 cases (1.8%) and vascular injury in 4 cases (1.5%). Postoperative complications occurred in 35 cases (12.2%), with chyle leakage being the most common, occurring in 18 cases (6.6%). Port site hernia was observed in three cases, and delayed intestinal injury was observed in one case; however, it was found that the number of cases was lower than in reports by other researchers. It is worth noting that para-aortic lymph node metastasis was observed in 29 cases (10.7%), and 11 cases (4%) had isolated para-aortic lymph node metastasis, which may have reaffirmed the importance of para-aortic lymph node dissection for early-stage endometrial cancer with an intermediate to high risk of recurrence. The NCCN guidelines also state that minimally invasive surgery is suitable for uterine endometrial cancer where the lesion remains within the uterine corpus, and this type of surgery is positioned as the standard procedure for early-stage uterine endometrial cancer [13]. On the other hand, there are no reports of any comprehensive research results on laparoscopic surgery for stage IV endometrial cancer that is suspected of having peritoneal dissemination or intestinal invasion. As with surgery for advanced ovarian cancer that has disseminated lesions, the prognosis for stage IV endometrial cancer is greatly affected by the size of the residual tumor, so debulking surgery is important. However, although the postoperative advanced stage IV rate in the RCTs mentioned above was 0–2%, the largest-scale LAP2 study found that conversion to open surgery occurred in 26% of cases with metastatic lesions. In other words, laparoscopic surgery for advanced endometrial cancer in cases where the disease has spread outside of the uterine corpus is not currently recommended for widespread use in general clinical practice and is thought to be a form of medical research. The NCCN guidelines also state that open surgery is necessary for such advanced cases [13].

8.3 Position of Robot-Assisted Surgery for Uterine Endometrial Cancer

In recent years, in the United States and Europe, robot-assisted surgery has become more widespread than open and laparoscopic surgery for uterine endometrial cancer (Fig. 8.3), with 33.6% of cases involving open surgery, 17.6% involving laparoscopic surgery, and 48.8% involving robot-assisted surgery [14]. In Japan, robot-assisted surgery for early-stage uterine endometrial cancer became covered by health insurance in 2018 and is currently being used. With robot-assisted surgery, 3D high-resolution images allow the surgeon to see the organ being operated on directly in front of them, giving them a sense of being able to touch it with their own eyes. Freely moving forceps and image stabilization functions enable intuitive forceps manipulation, thus making it possible to perform more delicate and stable surgical manipulation than with the limited forceps use in laparoscopic surgery. A meta-analysis of the usefulness of robot-assisted surgery found that it was significantly superior to open surgery in terms of perioperative complications, blood loss, blood transfusion rate, and hospital stay, and there was no difference in the number of lymph nodes removed; however, it tended to be longer in terms of operating time [15]. Compared to laparoscopic surgery, it was superior in terms of blood loss, length of hospital stay, intraoperative complications, and conversion to open surgery, and there was no significant difference in terms of operating time, blood transfusion rate, postoperative complications, or number of lymph nodes removed [15]. In terms of prognosis, early-stage endometrial cancer surgery using robot-assisted surgery is no different from open or laparoscopic surgery, and it is thought that the curability is the same [9]. In addition, in a Swedish prospective comparative study of robotic para-aortic lymphadenectomy for stage I and II uterine corpus cancer,

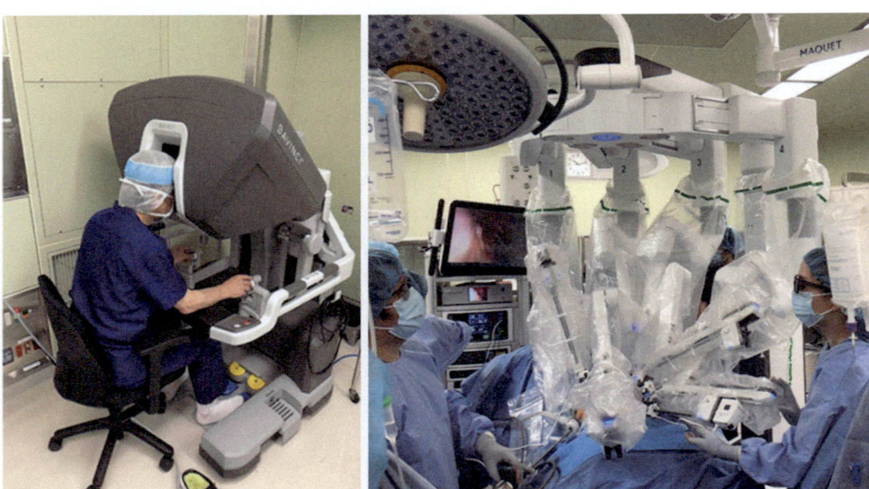

Fig. 8.3 da Vinci^{R} surgical system

which is considered to be at high risk of recurrence [16], the number of pelvic lymph nodes removed tended to be significantly lower in the robot-assisted surgery group ($p < 0.01$), at 22 ± 8.1, compared to 28 ± 10 in the open surgery group. However, there was no significant difference in the number of para-aortic lymph nodes removed, at 20.9 ± 9.6 in the robot-assisted surgery group compared to 22 ± 11 in the open surgery group, thus proving non-inferiority in the primary endpoint (para-aortic nodal yield). Similarly, there was no difference in perioperative complications, intraoperative blood loss, or hospital stay. In Japan, since April 2018, robot-assisted surgery has been covered by health insurance for early-stage uterine corpus cancer (corresponding to IA stage), as well as laparoscopic surgery, including pelvic lymph node dissection. However, overseas, robot-assisted surgery and laparoscopic surgery are not considered separately, and the conditions for application are judged to be the same as for laparoscopic surgery. Robot-assisted surgery is expected to become more widespread in the future, as it allows for more delicate surgical manipulation than laparoscopic surgery, owing to the use of high-resolution 3D images and surgical forceps that can be moved to any angle, therefore making surgical manipulation more intuitive.

8.4 Minimally Invasive Surgery for Obese Patients

Obesity is a risk factor for endometrial cancer, and many cases of endometrial cancer are experienced in obese patients. It is known that the risk of intraoperative complications is high in such patients undergoing surgery. A study of 1266 cases in Italy reported that the rate of lymph node dissection was significantly lower in cases of severe obesity with a BMI of over 40 and that the rate of conversion from laparoscopic surgery to open surgery was significantly higher, at 8.6%, in cases with a BMI of over 40 [15]. In a systematic review of open surgery and laparoscopic surgery, it was clear that obese cases had a higher overall complication rate than non-obese cases; there was also no significant difference between obese and non-obese cases in terms of intraoperative complications such as bladder injury, ureteral injury, and vascular injury. However, in open surgery, wound dehiscence occurred significantly more frequently in obese patients, and the rate of antibiotic administration due to postoperative infection tended to be higher [17]. The amount of intraoperative bleeding also tended to be significantly higher in obese patients, even in laparoscopic surgery, but was significantly lower in laparoscopic surgery than in open surgery. Therefore, it was concluded that laparoscopic surgery is more suitable for obese patients. In a study of the complication rates in 1087 cases of morbid obesity with a BMI of over 40 for uterine corpus cancer, the complication rates were 8% for robot-assisted surgery, 13% for laparoscopic surgery, and 23% for open surgery. In addition, it was reported that there was a 2.77-fold risk of complications for open surgery and a 1.92-fold risk of complications for laparoscopic surgery compared to robot-assisted surgery [18]. Additionally, the results of a systematic review of laparoscopic surgery and robot-assisted surgery [19] showed that organ/vascular injury (laparoscopic surgery, 3.5%; robot-assisted surgery, 1.2%), venous thromboembolism (laparoscopic surgery, 0.5%; robot-assisted surgery, 0.5%), and blood transfusions (laparoscopic surgery, 2.8%; robot-assisted surgery, 2.1%) were all lower in robot-assisted surgery, although there was no significant difference overall. When the conversion rate to open surgery was examined for both laparoscopic and robot-assisted surgery, the most common reason for conversion to open surgery was difficulty during surgery due to adhesions or visceral fat, and this was the same for both laparoscopic and robot-assisted surgery. However, there was no significant difference in the conversion rate to open surgery for a BMI of 30 or over, with 6.5% for laparoscopic surgery and 5.5% for robot-assisted surgery, and for a BMI of 40 or over, with 7.0% for laparoscopic surgery and 3.5% for robot-assisted surgery. The rate of conversion to open surgery due to difficulties in maintaining anesthesia was significantly lower for robot-assisted surgery (6%) than for laparoscopic surgery (31%). From the above, it can be considered that minimally invasive surgery, such as laparoscopic surgery or robot-assisted surgery, is a more suitable option than open surgery for obese patients with endometrial cancer, in terms of complications, and that robot-assisted surgery is more suitable than laparoscopic surgery if the surgeon is sufficiently skilled in the technique.

8.5 Future Directions

We have discussed the position of minimally invasive surgery for uterine endometrial cancer. Minimally invasive surgery, especially robot-assisted surgery, has many advantages and is sure to play a central role in this field in the near future. However, it is still a relatively new technology, and it is thought that there is a need to build up more evidence on the benefits and clinical outcomes of minimally invasive surgery for uterine endometrial cancer. In addition, although da Vinci^{R} had dominated the field of robot-assisted surgery for more than 10 years (Fig. 8.3), Hinotori^{TM} (Figs. 8.4

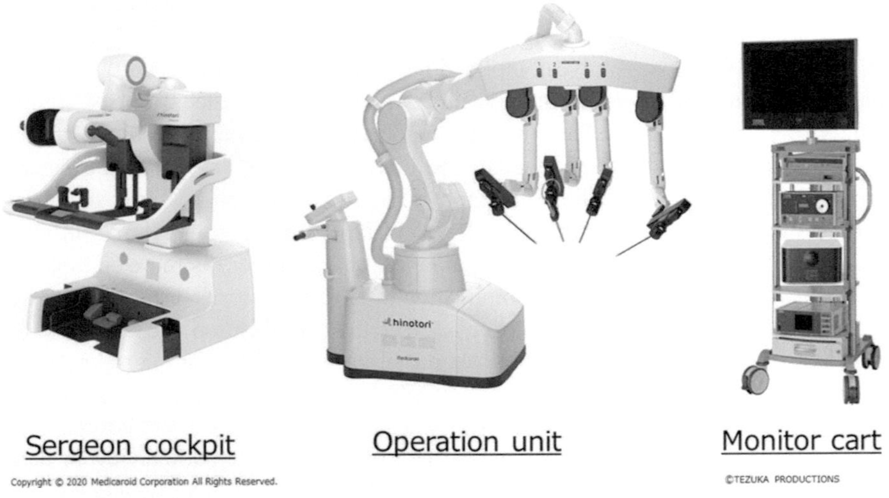

Fig. 8.4 Hinotori^{TM} surgical robot system

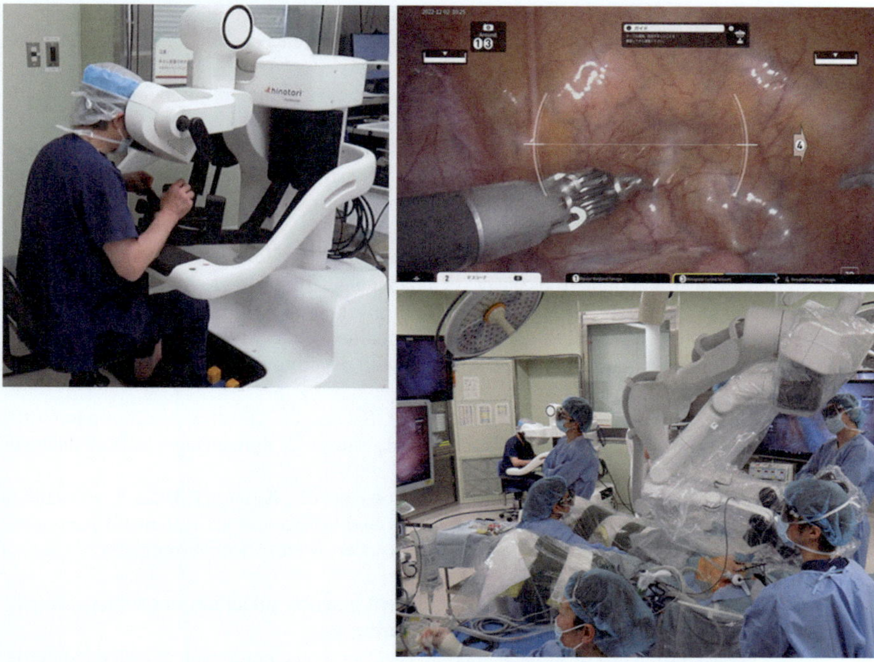

Fig. 8.5 View of robot-assisted surgery using the Hinotori™ surgical robot system

and 8.5), the Saroa surgical system, and Hugo have now been developed, and clinical use of these robots is now possible in Japan. In the future, we can expect to see the development and evolution of a variety of other new robotic surgical devices.

References

1. Yoshino K, Kurita T, Takahashi F, Nagase S, Board members of the 2022 Committee on Gynecologic Oncology of the Japan Society of Obstetrics and Gynecology. Annual report of the Committee on Gynecologic Oncology, the Japan Society of Obstetrics and Gynecology: Annual Patient Report for 2020 and Annual Treatment Report for 2015. J Obstet Gynaecol Res. 2023;49(11):2584–92.
2. Walker JL, Piedmonte MR, Spirtos NM, Eisenkop SM, Schlaerth JB, Mannel RS, et al. Laparoscopy compared with laparotomy for comprehensive surgical staging of uterine cancer: Gynecologic oncology group study LAP2. J Clin Oncol. 2009;27:5331–6.
3. Childers JM, Surwit EA. Combined laparoscopic and vaginal surgery for the management of two cases of stage I endometrial cancer. Gynecol Oncol. 1992;45(1):46–51.
4. Zullo F, Palomba S, Falbo A, Russo T, Mocciaro R, Tartaglia E, et al. Laparoscopic surgery vs laparotomy for early stage endometrial cancer: long-term data of a randomized controlled trial. Am J Obstet Gynecol. 2009;200:296 e1–9.
5. Janda M, Gebski V, Brand A, Hogg R, Jobling TW, Land R, et al. Quality of life after total laparoscopic hysterectomy versus total abdominal hysterectomy for stage I endometrial cancer (LACE): a randomised trial. Lancet Oncol. 2010;11:772–80.

6. Mourits MJ, Bijen CB, Arts HJ, ter Brugge HG, van der Sijde R, Paulsen L, et al. Safety of laparoscopy versus laparotomy in early-stage endometrial cancer: a randomised trial. Lancet Oncol. 2010;11:763–71.
7. He H, Zeng D, Ou H, Tang Y, Li J, Zhong H. Laparoscopic treatment of endometrial cancer: systematic review. J Minim Invasive Gynecol. 2013;20:413–23.
8. Walker JL, Piedmonte MR, Spirtos NM, Eisenkop SM, Schlaerth JB, Mannel RS, et al. Recurrence and survival after random assignment to laparoscopy versus laparotomy for comprehensive surgical staging of uterine cancer: Gynecologic oncology group LAP2 study. J Clin Oncol. 2012;30:695–700.
9. Wright JD, Burke WM, Tergas AI, Hou JY, et al. Comparative effectiveness of minimally invasive hysterectomy for endometrial cancer. J Clin Oncol. 2016;34:1087–96.
10. Terai Y, Tanaka T, Sasaki H, Kawaguchi H, Fujiwara S, Yoo S, Tanaka Y, Tsunetoh S, Kanemura M, Ohmichi M. Total laparoscopic modified radical hysterectomy with lymphadenectomy for endometrial cancer compared with laparotomy. J Obstet Gynaecol Res. 2014;40:570–5.
11. Todo Y, Kato H, Kaneuchi M, Watari H, Takeda M, Sakuragi N. Survival effect of para-aortic lymphadenectomy in endometrial cancer (SEPAL study): a retrospective cohort analysis. Lancet. 2010;375:1165–72.
12. Tanaka T, Terai Y, Hayashi S, Aoki D, Miki M, Kobayashi E, Kimura T, Baba T, Matsumura N, Ohmichi M. Comparison between laparoscopy and laparotomy in systematic para-aortic lymphadenectomy for patients with endometrial cancer: a retrospective multicenter study. J Gynecol Surg. 2017;33(3):105–10.
13. Uterine neoplasms (Version 3, 2025) NCCN clinical practice guidelines in oncology. https://www.nccn.org/professionals/physician_gls/pdf/uterine.pdf.
14. Bixel K, et al. Determinants of surgical approach and survival among women with endometrial carcinoma. J Minim Invasive Gynecol. 2022;29:219–30.
15. Park DA, Lee DH, Kim SW, Lee SH. Comparative safety and effectiveness of robot-assisted laparoscopic hysterectomy versus conventional laparoscopy and laparotomy for endometrial cancer. Eur J Surg Oncol. 2016;42:1303–14.
16. Salehi S, Åvall-Lundqvist E, Legerstam B, Carlson JW, Falconer H. Robot-assisted laparoscopy versus llaparotomy for infrarenal paraaortic lymphadenectomy in women with high-risk endometrial cancer: a randomized controlled trial. Eur J Cancer. 2017;79:81–9.
17. Uccella S, Bonzini M, Palomba S, et al. Impact of obesity on surgical treatment for endometrial cancer: a multicenter study comparing laparoscopy vs open surgery, with propensity-matched analysis. J Minim Invasive Gynecol. 2016;23:53–61.
18. Chan JK, et al. Robotic versus laparoscopic versus open surgery in morbidly obese endometrial cancer patients – a comparative analysis of total charges and complication rates. Gynecol Oncol. 2015;139:300–5.
19. Cusimano MC, Simpson AN, Dossa F, Liani V, Kaur Y, Acuna SA, Robertson D, Satkunaratnam A, Bernardini MQ, Ferguson SE, Baxter NN. Laparoscopic and robotic hysterectomy in endometrial cancer patients with obesity: a systematic review and meta-analysis of conversions and complications. Am J Obstet Gynecol. 2019;221(5):410–428.e19.

Immune Checkpoint Inhibitors in Endometrial Cancer: Current Status and Future Perspectives

9

Junzo Hamanishi

Abstract

Endometrial cancer (EC), the most common gynecological malignancy in developed countries, presents a significant challenge in its advanced or recurrent stages. While early-stage EC is often curable, advanced or recurrent EC has a poor prognosis, necessitating novel therapeutic strategies. In recent years, immune checkpoint inhibitors (ICIs) have emerged as a promising therapeutic strategy for EC, particularly in tumors harboring mismatch repair deficiency (MMRd) or high microsatellite instability (MSI-H). Subsequently, the combination of ICIs and lenvatinib, a multi-kinase inhibitor, has significantly improved survival of patients with previously treated recurrent EC, irrespective of MSI status. Besides, several phase III trials demonstrated the significant antitumor efficacy of combining ICIs plus first-line chemotherapy, followed by maintenance of the combination of ICIs with/without poly ADP ribose polymerase (PARP) inhibitors for both MMRd and mismatch repair proficient (MMRp) EC. This review summarizes the current landscape of ICIs in EC, including mechanisms of action, key clinical trials, emerging predictive biomarkers, real-world evidence, and ongoing research. Future perspectives focusing on combination therapies, biomarker-driven patient selection, and novel immunotherapeutic approaches are also discussed.

Keywords

Immune checkpoint inhibitor · PD-1 · PD-L1 · Mismatch repair (MMR) · Poly ADP ribose polymerase (PARP) inhibitor

J. Hamanishi (✉)
The Department of Obstetrics and Gynecology, National Hospital Organization Kyoto Medical Center, Kyoto, Japan
e-mail: jnkhmns@kuhp.kyoto-u.ac.jp

H. Watari, T. Mitamura (eds.), *Uterine Body Neoplasms*, Comprehensive Gynecology and Obstetrics, https://doi.org/10.1007/978-981-95-0946-1_9

9.1 Mechanisms of Action of Immune Checkpoint Inhibitors for Cancer

Tumor cells have acquired several ways to escape host immunity in the tumor microenvironment [1]. In the past two decades, several studies on cancer immune escape have revealed that one of the most important components of the underlying mechanisms is the immunosuppressive immune checkpoint signal via programmed cell death-1 (PD-1)/PD-1 ligand 1 or 2 (PD-L1 or PD-L2) in the tumor microenvironment. PD-1, discovered in 1992 [2] and expressed on T cells, negatively regulates the anti-tumor effects of T cells [3–5]; PD-L1 binds to PD-1 and inhibits T cell proliferation and cytokine production [6]. Under normal physiological conditions, PD-L1 is expressed in human tonsils, placental syncytia, monocytes, and lungs and has been shown in several studies to be involved in immune tolerance [7–9]. In addition, several preclinical studies have reported that inhibition of the PD-1-PD-L1 interaction enhances T cell responses and mediates anti-tumor activity [3–5, 8]. Furthermore, PD-L1 is expressed in a variety of human solid malignancies, including urothelial, gastrointestinal, lung, breast, melanoma, and ovarian cancers, as well as tumor-infiltrating immune cells in the tumor microenvironment [8–13]. Thus, to prove the principle that PD-1/PD-L1 blockade is effective against cancer, numerous anti-PD-1 or anti-PD-L1 antibodies have been clinically developed to treat human malignant tumors, with success in many kinds of malignant tumors [14].

9.2 PD-1 Signal in EC

Endometrial cancer (EC) cells have been shown to modulate the immune response via the PD-1 signal in the tumor microenvironment. Immunohistochemical analyses demonstrated that PD-L1 has been reported to be expressed in tumor specimens from approximately 10% to 80% of uterine cancer patients [5, 15–17]. Besides, subtypes of EC with high tumor mutational burden such as DNA polymerase ε (*POLE*) mutated or high microsatellite instability (MSI-H) exhibit many tumor neoantigens and induce tumor-infiltrating lymphocytes with upregulation of immune checkpoint signals [18–24].

Among several solid tumors, EC has the highest prevalence of mismatch repair deficiency (MMRd)/MSI-H status; 13–30% of ECs are reported to be MSI-H or MMRd status [25]. Therefore, immune check point inhibitor (ICI) monotherapy with/without combination strategies with other drugs such as targeted therapies or systemic chemotherapies has emerged as a promising approach to enhance the anti-tumor immune response in advanced or recurrent EC.

9.3 Key Clinical Trials (Monotherapy) in EC

Given the presence of immune dysregulation in EC as described, immune check point inhibitors (ICIs) have been explored as a therapeutic mechanism, both as monotherapy and in combination with cytotoxic chemotherapy, other immunotherapies, or targeted agents. The molecular classifications described by The Cancer Genome Atlas are increasingly being used in clinical research to identify patients with EC who may have a more robust response to immunotherapy. Given that MSI status is a marker for response to some ICIs, it is likely attributable to its high mutational burden and generation of neoantigens [26].

9.3.1 Pembrolizumab

In a phase Ib (KEYNOTE-028) study of pembrolizumab in advanced PD-L1-positive solid tumors, 24 patients with recurrent metastatic EC had an objective response rate (ORR) of 13% (three patients with partial response) and an additional 13% (three patients) achieved stable disease. Of the three patients achieving a partial response, one had non-MSI-H EC with a *POLE* mutation, one had non-MSI-H EC, and the other had an unknown status [27]. Prior to FDA approval, the first evidence for clinical activity of immunotherapy in advanced EC was derived from a phase II study of 41 patients with heavily pretreated metastatic carcinoma with or without MMRd and also included two patients with endometrial carcinomas, and these patients achieved a partial response [28].

The antitumor efficacy of pembrolizumab monotherapy was further investigated in a phase II trial across 12 MMRd tumor types [29]. The ORR of 53% including 21% with a complete response (CR) was observed among the 86 patients enrolled. The EC cohort demonstrated a similar ORR of 53%. Another phase II KEYNOTE-158 study across 27 advanced MMRd/MSI-H solid tumors also yielded consistent positive results [30]. This study included 49 patients with EC and had an overall response rate of 57%, including 16% with a CR. These results led to the FDA approval of pembrolizumab for MMRd/MSI-H solid tumors including EC.

9.3.2 Dostarlimab

Preliminary data from a phase I GARNET study of the anti-PD-1 antibody dostarlimab demonstrated an ORR of 49% in MSI-H advanced EC compared to an ORR of 20% in MSS tumors. The median duration of response (DOR) was not reached, with approximately half of all responders receiving treatment for less than 1 year. Dostarlimab was well tolerated, with less than 6% of patients experiencing grade 3 or higher immune-related adverse events [31]. These results led to the FDA approval of dostarlimab for MMRd EC. Thereafter, in this trial, a total of 143 patients with MMRd/MSI-H EC and 156 patients with MMRp/MSS were evaluated, and ORR was 45.5% and 15.4% for MMRd/MSI-H and MMRp/MSS, respectively. The median DOR for MMRd/MSI-H was not met (median follow-up, 27.6 months); the median DOR for MMRp/MSS was 19.4 months. The ORRs by combined positive score (CPS) ≥1 status were 54.9% and 21.7% for MMRd/MSI-H and MMRp/MSS, respectively. The safety profile of dostarlimab was consistent with previous reports. These data demonstrated the durable antitumor activity and safety of dostarlimab in patients with MMRd/MSI-H EC. Biomarkers associated with EC may identify patients likely to respond to dostarlimab [32], and now dostarlimab is under investigation in a phase III trial in the first-line setting, in combination with carboplatin and paclitaxel chemotherapy with/without the PARP inhibitor niraparib (RUBY trial; NCT03981796).

9.3.3 Other ICIs

A phase II trial of the anti-PD-L1 antibody durvalumab (PHAEDRA/ANZGOG1601 trial) has shown that the ORR in the MMRd cohort was 47% (17/36 patients), including 6 CR and 11 PR vs 3% in the MMRp cohort (1/35 patients). Median progression-free survival (PFS) was 8.3 months in the MMRd cohort vs 1.8 months in the MMRp cohort. The 12-month overall survival (OS) rate was 71% in MMRd patients vs 51% in MMRp patients, with median OS not reached for MMRd patients vs 12 months for MMRp patients. This study showed promising activity and acceptable safety in EC with MMRd regardless of prior lines of chemotherapy, but activity was limited with MMRp patients [33].

Another phase II trial of the anti-PD-L1 antibody avelumab has shown ORRs of 26.7% in MMRd EC and 6.25% in MMRp tumors, respectively [34]. Thereafter, to assess the potential positive outcomes and safety of the combination of the PARP inhibitor talazoparib and the anti-PD-L1 antibody avelumab in recurrent MMRp EC, a single-arm, two-stage, phase 2 study was conducted. Thirty-five patients received protocol therapy, with 9 (25.7%) experiencing clinical benefit after meeting at least one of the two coprimary endpoints. Four patients (11.4%) exhibited confirmed ORRs, and 8 (22.9%) survived progression free at 6 months. Tumors with homologous recombination repair alterations were associated with clinical benefit from treatment with avelumab and talazoparib [35].

9.4 Key Clinical Trials (Combination) in EC

Based on the promising results of the abovementioned clinical trials of a single-agent ICI, a number of phase III trials have been developed in anticipation of an add-on maintenance effect to the first-line standard chemotherapy, and a number of regulatory approvals have already been granted. This section outlines the individual clinical trials in bullet points (Table 9.1, Fig. 9.1).

Table 9.1 Phase III trials of ICI + systemic chemotherapy for advanced or recurrent endometrial cancer

Trial	NRG GY018 [a,b]	RUBY 1 [c,d,e]	AtTEnd [f]	DUO-E [g,h]	ENGOT-EN6/RUBY 2 [i]
n	816	494	671	718	291
Treatment vs Control	Pembrolizumab + CT pbo + CT†	Dostarlimab + CT pbo + CT†	Atezolizumab + CT pbo + CT†	Durvalumab + CT ± olaparib pbo + CT†	Dostarlimab + CT+ niraparib pbo + CT†
Patient population	No carcinosarcoma, MMR testing by IHC	No potential for cure by RT and/or surgery; no prior systemic tx or recurrence	With residual disease either measurable or evaluable, or inoperable stage III-IV endometrial carcinoma/carcinosarcoma	Newly diagnosed, measurable stage III/IV (FIGO 2009) or recurrent endometrial (no sarcoma); previous 1L systemic tx; no previous PARPi or IO	No potential for cure by RT and/or surgery; no prior systemic tx or recurrence
previous systemic CT	PFI ≥12 mo	PFI ≥6 mo	PFI ≥12 mo	PFI ≥12 mo	PFI ≥6 mo
PFS	**Median PFS, mo**	**2-yr PFS, %**	**Median PFS, mo**	**Median PFS, mo**	**1-yr PFS, %**
	dMMR: NR vs 8.3 (HR: 0.34)	dMMR: 61.4 vs 15.7	dMMR: NE vs 6.9	dMMR: 31.8‡ vs NE vs 7.0	dMMR: 64.4 vs 40.8 (HR 0.48)
	pMMR: 13.1 vs 8.7 (HR: 0.57)	pMMR: 28.4 vs 18.8	pMMR: 9.5 vs 9.2	pMMR: 15.0‡ vs 9.9 vs 9.7	pMMR: 54.7 vs 31.1 (0.63)
OS	**Median OS, mo**	**Median OS, mo**	**Median OS, mo**	**Median OS, mo**	x
	Overall: 27.96 vs 27.37	Overall: 44.6 vs 28.2 (HR: 0.69)	x	ITT: NE‡ vs NE vs 25.9	x
	dMMR: NR vs NR (HR: 0.55)	dMMR: NE vs 31.4 (HR: 0.32)	dMMR: NE vs 25.7	x	x
	pMMR: 8.8 vs 8.4 (HR: 0.79)	pMMR: 34.0 vs 27.0 (HR: 0.79)	pMMR: 31.5 vs 28.6	x	x
	x	**2-yr OS, %**	x	x	x
	x	dMMR: 82.8 vs 57.5	x	x	x
	x	pMMR: 66.5 vs 53.2	x	x	x
ID	KEYNOTE 868	ENGOT-EN6-NSGO/ GOG-3031	EudraCT 2018-001072-37	ENGOT-EN10/ GOG 3041	ENGOT-EN6-NSGO/ GOG-3031
NCT	3914612	3981796	3603184	4269200	3981796

j: *CT* chemotherapy, *dMMR* mismatch repair deficient *IO* immuno-oncology, *ITT* intention-to-treat population, *NE* not evaluable, *OS* overall survival, *pbo* placebo, *PFI* platinum-free interval, *PFS* progression-free survival, *pMMR* mismatch repair proficient, *RT* radiation therapy, *tx* treatment

[a]Eskander. NEJM. 2023; 388:2159

[b]Eskander. SGO 2024

[c]Mirza. NEJM. 2023; 388:2145

[d]Dostarlimab PI

[e]Powell. SGO 2024

[f]Colombo. ESMO 2023. Abstr LBA40

[g]Westin. J Clin Oncol. 2023; 42:283

[h]Westin. J Clin Oncol. 2024; 42:283

[i]Mirza SGO 2024

9.4.1 KEYNOTE 775/Study 309

The KEYNOTE-775/Study 309 trial evaluated the multi-kinase inhibitor lenvatinib plus pembrolizumab versus the physician's choice chemotherapy (TPC) in previously treated advanced EC ($N = 827$). Lenvatinib inhibits several kinases involved in tumor angiogenesis, while pembrolizumab is an anti-PD-1 antibody. The trial sought to address the poor prognosis of advanced EC, which has a 5-year survival

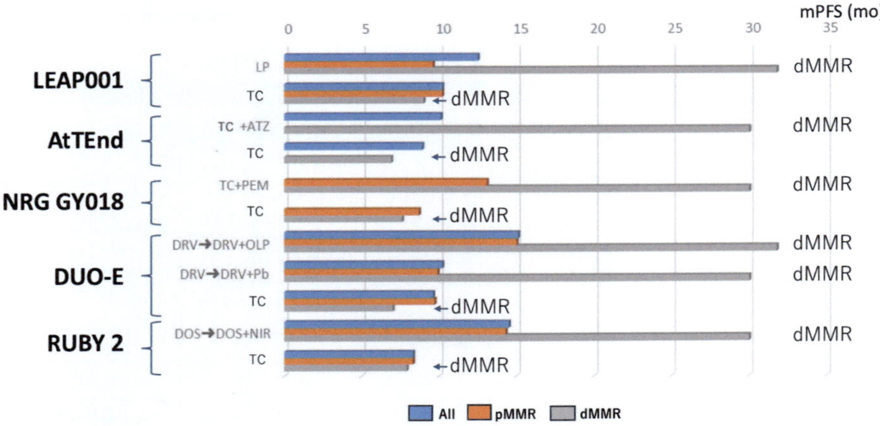

Fig. 9.1 PFS of first-line chemotherapy with ICI for MMRd/MMRp advanced or recurrent EC. LP lenvatinib + pembrolizumab, TC paclitaxel + carboplatin, ATZ atezolizumab, PEM pembrolizumab, DRV durvalumab, OLP olaparib, Pb placebo, DOS dostarlimab, NIR niraparib. (Reference: Marth. SGO2024. Colombo ESMO2023. Powell. SGO2024. Eskander ESMO2023. Mirza. SGO2024)

rate of approximately 17%. Results showed significant improvements in the lenvatinib/pembrolizumab arm compared to TPC. In the MMRp subgroup, OS was significantly improved (HR: 0.70, 95% CI: 0.58–0.83), and PFS was also improved (HR: 0.60, 95% CI: 0.55–0.77), compared to chemotherapy. In the overall population, both OS and PFS were improved (OS, HR: 0.65, 95% CI: 0.55–0.77; PFS, HR: 0.56, 95% CI: 0.48–0.66). ORRs were also higher in the lenvatinib/pembrolizumab arm at 32.4% in the MMRp subgroup and 33.8% overall, compared to 15.1% and 14.7%, respectively, in the TPC arm. While nearly all patients in both arms experienced treatment-emergent adverse events, no new safety signals were identified. The study concluded that lenvatinib plus pembrolizumab demonstrated superior efficacy with a manageable safety profile compared to TPC in previously treated advanced EC [36]. These results led to the FDA approval of lenvatinib plus pembrolizumab therapy for recurrent EC.

9.4.2 NRG GY018/Keynote 868

The NRG-GY018 study was designed to evaluate the efficacy of pembrolizumab plus paclitaxel plus carboplatin as initial treatment for measurable stage III, IVA, and IVB or recurrent uterine cancer ($N = 816$ patients), with six cycles of pembrolizumab plus paclitaxel plus carboplatin for one cycle of 3 weeks as first-line treatment, followed by one cycle of 6 weeks as maintenance treatment with pembrolizumab. Patients were randomized in a 1:1 ratio to receive pembrolizumab plus carboplatin plus paclitaxel for 6 cycles followed by pembrolizumab every 6 weeks for up to 14 cycles, or placebo plus carboplatin plus paclitaxel for 6 cycles followed by placebo maintenance. The primary endpoint was PFS. Patients with

MMRd and patients without mismatch repair mechanism deficiency (MMRp) were analyzed separately. The results of the study showed that the 12-month PFS rate in the MMRd group was 74% with pembrolizumab compared to 38% with placebo, with a 70% reduced risk of disease progression or death with pembrolizumab (HR: 0.30, 95% CI: 0.19–0.48, $P < 0.001$). PFS in the MMRp patient group was 8.7 months in the placebo group compared to 13.1 months in the pembrolizumab group, with a 46% reduction in risk of disease progression or death with the pembrolizumab combination (HR: 0.54, 95% CI: 0.41–0.71, $P < 0.001$). These results from the NRG-GY018 study show that the addition of the anti-PD-1 antibody drug pembrolizumab to standard chemotherapy as the first-line treatment for patients with recurrent/advanced uterine cancer resulted in a statistically significant improvement in PFS [37]. These results led to the FDA approval of pembrolizumab plus first-line chemotherapy for MMRd/p advanced and recurrent EC.

9.4.3 RUBY part1/ENGOT-EN6-NSGO/GOG-3031

The RUBY study was divided into PART 1 and PART 2. The primary endpoints of PFS and OS were statistically analyzed in all patients (ITT) and patients with microsatellite instability (MSI-H) or mismatch repair defects (MMRd), respectively, as assessed by the primary investigator. Secondary endpoints include progression-free survival (PFS), objective response rate (ORR), duration of response (DOR), and disease control rate (DCR) as determined by an independent evaluation committee [38].

The RUBY part1 study evaluated dostarlimab plus carboplatin/paclitaxel, followed by dostarlimab maintenance versus placebo plus chemotherapy, followed by placebo in patients with advanced/recurrent EC. The primary endpoint, median overall survival (OS), was 44.6 months in the dostarlimab group compared to 28.3 months in the placebo group, representing a 31% improvement in risk of death (HR: 0.69, 95% CI: 0.54–0.89, $P = 0.002$). In the MSI-H/MMRd subgroup, median OS was not reached in the dostarlimab group but was 31.4 months in the placebo group, with a 68% reduction in risk of death. In the MSS/MMRp group, median OS was 34.0 months in the dostarlimab group and 27.0 months in the placebo group, with a 21% reduction in risk of death (HR: 0.78, 95% CI: 0.60–1.00).

The 2-year OS rate was 82.8% in the dostarlimab group versus 57.5% in the placebo group, and the 3-year OS rate was 78.0% in the dostarlimab group versus 46.0% in the placebo group. The median OS in the microsatellite stability (MSS) or mismatch repair defect-free (MMRp) patient group was 27.0 months in the placebo group compared with 34.0 months in the dostarlimab group. The risk of death was reduced by 21% (HR: 0.79, 95% CI: 0.60–1.04) in the dostarlimab group. In addition, the 2-year OS rate was 66.5% in the dostarlimab group compared with 53.2% in the placebo group, and the 3-year OS rate was 48.6% in the dostarlimab group compared with 41.9% in the placebo group.

Grade 3 or higher treatment-related adverse events (TRAEs) occurred in 72.2% of the dostarlimab group and 60.3% of the placebo group. Serious TRAEs were observed in 39.8% and 28.0%, respectively. Immune-related adverse events occurred in 40.7% and 16.9%, respectively.

The study concluded that dostarlimab plus carboplatin/paclitaxel is a potential new standard of care for patients with primary advanced or recurrent EC, irrespective of mismatch repair status [38]. These results led to the FDA approval of dostarlimab plus first-line chemotherapy for MMRd advanced and recurrent EC.

9.4.4 AtTEnd

The AtTEnd trial randomized untreated advanced or recurrent uterine cancer patients to receive either carboplatin + paclitaxel + atezolizumab or carboplatin + paclitaxel + placebo as initial therapy. In the MMRd population, the PFS was 6.9 months in the chemotherapy arm compared to not reached in the atezolizumab arm, representing a 64% risk reduction in disease progression or death (HR: 0.36, 95% CI: 0.23–0.57, $P = 0.0005$). In the overall population, the PFS was 8.9 months in the chemotherapy arm compared to 10.1 months in the atezolizumab arm, with a 26% reduction in risk of disease progression or death (HR: 0. 69, 95% CI: 0.61–0.91, $P = 0.022$). Overall survival (OS) in the overall population was 30.2 months in the chemotherapy arm compared to 38.7 months in the atezolizumab arm, showing an 18% reduction in the risk of death (HR: 0.82, 95% CI: 0.63–1.07, $P = 0.048$). On the other hand, common grade 3 or 4 adverse events included neutropenia, anemia, and severe adverse events, with neutropenia being more common in the chemotherapy arm. Serious treatment-related adverse events were more common in the chemotherapy group compared to the atezolizumab combination group [39].

These results showed that adding atezolizumab to paclitaxel plus carboplatin as first-line therapy for patients with advanced or recurrent uterine cancer improves PFS, particularly in the MMRd population.

9.4.5 LEAP001/ENGOT-en9/MK-7902-001

This study evaluated lenvatinib plus pembrolizumab versus paclitaxel plus carboplatin as first-line treatment for advanced EC ($N = 842$). In the MMRp subgroup, median PFS was similar between the lenvatinib/pembrolizumab (9.6 months) and chemotherapy (10.2 months) arms (HR 0.99). In the overall population, median PFS was slightly longer with lenvatinib/pembrolizumab (12.5 months) compared to chemotherapy (10.2 months) (HR: 0.99, 95% CI: 0.82–1.21). Median OS in the MMRp subgroup was also similar (lenvatinib/pembrolizumab 30.9 months vs chemotherapy 29.4 months) (HR: 0.91, 95% CI: 0.76–1.09). Overall survival in the overall population was numerically longer with lenvatinib/pembrolizumab (37.7 months) vs chemotherapy (32.1 months). On the other hand, the safety profile of the paclitaxel plus carboplatin arm showed a 67% incidence of grade 3 or higher treatment-related adverse events (TRAEs), compared to 79% in the lenvatinib/pembrolizumab arm. The study concluded that lenvatinib plus pembrolizumab did not demonstrate a statistically significant improvement in PFS or OS as first-line treatment for advanced EC [40].

9.4.6 DUO-E/GOG 3041/ENGOT-EN10

The DUO-E trial evaluated the efficacy and safety of durvalumab plus platinum-based chemotherapy (carboplatin and paclitaxel) as first-line treatment, followed by durvalumab ± olaparib as maintenance treatment, compared with chemotherapy alone in patients with newly diagnosed advanced or recurrent uterine cancer. In the intention-to-treat population, statistically significant PFS benefit was observed in the durvalumab (HR: 0.71, 95% CI: 0.57–0.89; $P = 0.003$) and durvalumab + olaparib arms (HR: 0.55, 95% CI: 0.43–0.69; $P < 0.0001$) versus control. Prespecified, exploratory subgroup analyses showed PFS benefit in MMRd (HR [durvalumab vs control]: 0.42, 95% CI: 0.22–0.80]; HR [durvalumab + olaparib vs control]: 0.41, 95% CI: 0.21–0.75) and MMRp subgroups (HR [durvalumab vs control]: 0.77, 95% CI: 0.60–0.97]; HR [durvalumab + olaparib vs control]: 0.57, 95% CI: 0.44–0.73); and in PD-L1-positive subgroups (HR [durvalumab vs control]: 0.63, 95% CI: 0.48–0.83; HR [durvalumab + olaparib vs control]: 0.42, 95% CI: 0.31–0.57). Interim overall survival results (maturity approximately 28%) were supportive of the primary outcomes (durvalumab vs control, HR: 0.77, 95% CI: 0.56–1.07]; $P = 0.120$; durvalumab + olaparib vs control, HR: 0.59, 95% CI: 0.42–0.83], $P = 0.003$). In addition, the safety profile was generally manageable and well tolerated for both durvalumab plus chemotherapy and durvalumab plus olaparib, and generally consistent with the known profiles of the individual agents. This is the first approval for the combination of an immune checkpoint inhibitor with chemotherapy as first-line treatment and an immune checkpoint inhibitor with a PARP inhibitor for maintenance therapy in advanced or recurrent uterine cancer [41]. These results led to the FDA approval of durvalumab plus first-line carboplatin–paclitaxel followed by single-agent durvalumab for adult patients with primary advanced or recurrent MMRd endometrial cancer; in the European Union, following a positive EMA/CHMP opinion, the European Commission approved durvalumab plus first-line chemotherapy followed by durvalumab plus olaparib maintenance in MMRp disease and durvalumab maintenance in MMRd disease.

9.4.7 RUBY part2/ ENGOT-EN6-NSGO/GOG-3031

The RUBY part 2 study evaluated dostarlimab plus chemotherapy, followed by dostarlimab plus niraparib maintenance versus placebo plus chemotherapy, followed by placebo in stage III/IV or first recurrent EC of all histological subtypes except sarcoma [42]. In the overall population (median follow-up 21.7 months), median PFS was 14.5 months with dostarlimab vs 8.3 months with placebo (HR 0.60). The 12-month PFS rates were 57.0% vs 13.7%, respectively. In the MMRp/MSS population (median follow-up 21.8 months), median PFS was 14.3 months vs

5.3 months (HR 0.63). The 12-month PFS rates were 54.7% vs 31.1%, respectively. In the MMRd/MSI-H population (median follow-up 21.2 months), median PFS was not evaluable vs 7.9 months (HR 0.48). The 12-month PFS rates were 64.4% vs 40.8%, respectively. Further analysis showed PFS benefit with dostarlimab across molecular/histology subgroups (HR 0.59). Patients with *TP53*-mutated disease benefited most (HR 0.29). Exploratory PFS analyses favored dostarlimab regardless of PD-L1, BRCA, or HRR status (HR 0.59). Similar results were seen in the MMRp/MSS population (HR 0.62).

Regarding the safety data in the dostarlimab arm, 99.5% of patients experienced some of TEAEs (treatment-emergent adverse events), and 96.3% reported treatment-related TEAEs. Grade 3 or higher TEAEs were observed in 84.8% of patients, and treatment-related TEAEs of grade 3 or higher in 70.7%. Serious TEAEs were reported in 44% of patients, and treatment-related serious TEAEs in 23.5%. Some forms of dostarlimab-related AEs were seen in 36.6% of patients. TEAEs led to treatment discontinuation in 36.6% and death in 2.1% of patients. In the placebo arm, all evaluable patients experienced some form of TEAEs, 97.9% of which were treatment related. Grade 3 or higher TEAEs were observed in 49.0% of patients, and treatment-related TEAEs of grade 3 or higher in 36.5%. Serious TEAEs were reported in 19.8% of patients, and treatment-related serious TEAEs in 9.4%. Some forms of immune-related adverse effects (irAEs) were seen in 6.3% of patients. TEAEs led to treatment discontinuation in 13.5% of patients. No deaths were observed.

These outcomes demonstrate a potential role for PARP inhibitor niraparib maintenance in patients receiving dostarlimab plus chemotherapy, particularly in the MMRp population.

9.5 Ongoing Clinical Trials

Multiple ongoing studies aim to optimize the use of ICIs in EC, either in combination with other targeted agents (e.g., PARP inhibitors, anti-angiogenic drugs) or moving ICIs to earlier treatment lines (neoadjuvant or adjuvant settings). Several ongoing phase III trials are investigating immunotherapy's role in advanced/recurrent EC. The KEYNOTE-C93/ENGOT-en15/GOG-3064 trial is comparing pembrolizumab monotherapy vs chemotherapy in MMRd advanced/recurrent EC (NCT05173987). The DOMENICA/GINECO-EN105b/ENGOT-en13 trial is evaluating dostarlimab vs chemotherapy in MMRd/MSI-H advanced/recurrent EC, allowing crossover to dostarlimab at progression (NCT05201547). The MMRd-GREEN trial in RAINBOW program is studying adjuvant durvalumab plus radiotherapy vs radiotherapy alone in stage II-III MMRd EC (NCT05255653) [43–45].

Besides, new concepts of early phase trials are focused on IL-2 immunotherapy. One study evaluates IL-2 superkine MDNA11 with or without ICIs for solid tumors including EC (NCT05086692) [46]. The AFT-50 EndoMAP trial evaluates the efficacy and safety of biomarker (FoundationOne CDx NGS assay)-defined targeted agents for recurrent or persistent ECs (such as talazoparib for tumors with genomic loss of heterozygosity, trastuzumab emtansine for tumors with ERBB2/HER2 mutated or amplification, tiragolumab for MSI-H and/or TMB-H tumors, and ipatasertib for PIK3CA/PTEN/AKT1-altered tumors (NCT04486352)) [47]. The XmAb20717–05 trial evaluates a bispecific antibody, vudalimab, that simultaneously targets dual immune checkpoint receptors PD-1 and CTLA-4 and is designed for advanced gynecologic cancers including advanced EC [48]. The MK-6482-016 trial evaluates the efficacy and safety of the triplet regimens with pembrolizumab + lenvatinib with/without a second-generation small-molecule HIF-2α inhibitor belzutifan for advanced solid tumors including EC (NCT04976634) [49]. These trials aim to refine patient selection via biomarkers and enhance antitumor responses through immunotherapy-based combinations.

9.6 Biomarker Exploration of ICI Therapy for EC

There is considerable interest in biomarkers that can identify EC patients with a greater likelihood of response to treatments with ICI therapies (monotherapy/combination). Several analyses have evaluated the association between biomarkers and the response to ICI therapies, especially for MMRd/MSI-H and TMB-H among several types of malignant tumors including EC. Recent reports focus on T cell-inflamed gene expression profile (T cell-inflamed GEP) which is comprised of 18 genes that are indicative of a T cell-activated tumor microenvironment (TME). High T cell-inflamed GEP scores have demonstrated positive associations with response to pembrolizumab monotherapy across 20 cohorts of clinical trials for 20 types of solid tumors, including patients with EC [50, 51]. Additionally, other reflective factors of the TME and tumor biology include immune suppressive cells such as myeloid-derived suppressor cells (MDSC), angiogenesis, and stroma/epithelial-mesenchymal transition (EMT)/transforming growth factor β (TGFβ) sequence [52]. However, in KEYNOTE-146, T cell-inflamed GEP, angiogenesis, and EMT-related scores were not predictive biomarkers for the antitumor effect of the combination therapy of pembrolizumab plus lenvatinib [53].

From the results of all phase III trials of first-line systemic chemotherapy with ICI, RUBY part1, AtTEnd, and with ICI + PARP inhibitor, DUO-E and RUBY part2, the combination of ICI to paclitaxel plus carboplatin significantly prolonged patients' survival of MMRd ECs (Fig. 9.1). However, the AtTEnd trial demonstrated the negative result for MMRp endometrial patients, while other trials showed favorable PFS. Besides, subgroup analyses of DUO-E and RUBY part1 demonstrated that *TP53* mutation and/or serous histological subtypes tend to be more effective than other subtypes [54, 55].

9.7 Future Directions

ICIs have transformed the treatment paradigm of advanced and recurrent EC. Notably, patients with MMRd/MSI-H disease derive substantial benefit from anti-PD-1/PD-L1 monotherapies or in combination with standard chemotherapy. Meanwhile, novel combination regimens such as ICI plus multi-tyrosine kinase inhibitor or first-line chemotherapy plus ICI with/without PARP inhibitor have extended immunotherapy benefits to both MMRd/MSI-H and MMRp/MSS populations. Despite these advances, challenges remain, including the need to refine patient selection, develop effective combinations to overcome resistance, and manage immune-related toxicities.

Based on the results of the aforementioned phase III clinical trials, it is anticipated that standard treatment plus ICI will be progressively covered by insurance as a first-line treatment for patients with more advanced or recurrent EC, thus leading to a new therapeutic revolution for uterine cancer. Ongoing clinical trials and translational research hold promise for further improving outcomes, advancing the shift of ICIs from later lines of therapy into the frontline setting and achieving a more personalized, biomarker-driven approach to EC.

However, in actual clinical practice, it is not clearly defined which first-line treatment regimen and according to which criteria (e.g., MMR, HRD, PD-L1) should be selected for advanced or recurrent cases. A number of issues have also emerged, such as how to proceed with second-line treatment after first-line treatment (maintenance therapy) with ICI. In addition, it is necessary to consider the pros and cons of conversion surgery and salvage surgery using ICI-containing first-line treatment for inoperable cases, which has already started in colorectal cancer, gastric cancer, and other cancers [56–58]. Therefore, the Japanese Gynecologic Oncology Group (JGOG) is currently building real-world evidence through a prospective observational study (JGOG 2055s/PEACH study) of patients with advanced or recurrent uterine cancer previously treated with chemotherapy and requiring drug treatment, as well as a reverse translational study (JGOG 2055s-TR1/PEACH-TR study) using tumor/blood samples from the JGOG 2055s study [59]. The results of these studies are being used to clarify the actual status of uterine cancer pharmacotherapy in clinical practice, to identify differences in efficacy and side effects between regions and ethnic groups, and to search for biomarkers of these differences, with the aim of implementing more precise uterine cancer pharmacotherapy, thereby extending patient survival further.

Acknowledgements None.

Declarations **Conflict of Interest** JH is funded by Sumitomo Pharma, Kinopharma, Chugai, and Ono and receives lecture fee from MSD and Eisai.

Funding This work was supported by the Japan Society for the Promotion of Science (JSPS) through Grants-in-Aid for Scientific Research (grant number 21 K09541) to Junzo Hamanishi.

Data Availability My manuscript has no associated data.

References

1. Dunn GP, Bruce AT, Ikeda H, Old LJ, Schreiber RD. Cancer immunoediting: from immunosurveillance to tumor escape. Nat Immunol. 2002;3(11):991–8. https://doi.org/10.1038/ni1102-991.
2. Ishida Y, Agata Y, Shibahara K, Honjo T. Induced expression of PD-1, a novel member of the immunoglobulin gene superfamily, upon programmed cell death. EMBO J. 1992;11(11):3887–95. https://doi.org/10.1002/j.1460-2075.1992.tb05481.x.
3. Iwai Y, Terawaki S, Honjo T. PD-1 blockade inhibits hematogenous spread of poorly immunogenic tumor cells by enhanced recruitment of effector T cells. Int Immunol. 2005;17(2):133–44. https://doi.org/10.1093/intimm/dxh194.
4. Freeman GJ, Long AJ, Iwai Y, Bourque K, Chernova T, Nishimura H, Fitz LJ, Malenkovich N, Okazaki T, Byrne MC, Horton HF, Fouser L, Carter L, Ling V, Bowman MR, Carreno BM, Collins M, Wood CR, Honjo T. Engagement of the PD-1 immunoinhibitory receptor by a novel B7 family member leads to negative regulation of lymphocyte activation. J Exp Med. 2000;192(7):1027–34. https://doi.org/10.1084/jem.192.7.1027.
5. Keir ME, Butte MJ, Freeman GJ, Sharpe AH. PD-1 and its ligands in tolerance and immunity. Annu Rev Immunol. 2008;26:677–704. https://doi.org/10.1146/annurev.immunol.26.021607.090331.
6. Hirano F, Kaneko K, Tamura H, Dong H, Wang S, Ichikawa M, Rietz C, Flies DB, Lau JS, Zhu G, Tamada K, Chen L. Blockade of B7-H1 and PD-1 by monoclonal antibodies potentiates cancer therapeutic immunity. Cancer Res. 2005;65(3):1089–96.
7. Okazaki T, Honjo T. PD-1 and PD-1 ligands: from discovery to clinical application. Int Immunol. 2007;19(7):813–24. https://doi.org/10.1093/intimm/dxm057.
8. Hamanishi J, Mandai M, Konishi I. Immune checkpoint inhibition in ovarian cancer. Int Immunol. 2016;28(7):339–48. https://doi.org/10.1093/intimm/dxw020.
9. Hamanishi J, Mandai M, Iwasaki M, Okazaki T, Tanaka Y, Yamaguchi K, Higuchi T, Yagi H, Takakura K, Minato N, Honjo T, Fujii S. Programmed cell death 1 ligand 1 and tumor-infiltrating CD8+ T lymphocytes are prognostic factors of human ovarian cancer. Proc Natl Acad Sci USA. 2007;104(9):3360–5. https://doi.org/10.1073/pnas.0611533104.
10. Ghebeh H, Mohammed S, Al-Omair A, Qattan A, Lehe C, Al-Qudaihi G, Elkum N, Alshabanah M, Bin Amer S, Tulbah A, Ajarim D, Al-Tweigeri T, Dermime S. The B7-H1 (PD-L1) T lymphocyte-inhibitory molecule is expressed in breast cancer patients with infiltrating ductal carcinoma: correlation with important high-risk prognostic factors. Neoplasia. 2006;8(3):190–8. https://doi.org/10.1593/neo.05733.
11. Konishi J, Yamazaki K, Azuma M, Kinoshita I, Dosaka-Akita H, Nishimura M. B7-H1 expression on non-small cell lung cancer cells and its relationship with tumor-infiltrating lymphocytes and their PD-1 expression. Clin Cancer Res. 2004;10(15):5094–100. https://doi.org/10.1158/1078-0432.CCR-04-0428.
12. Nomi T, Sho M, Akahori T, Hamada K, Kubo A, Kanehiro H, Nakamura S, Enomoto K, Yagita H, Azuma M, Nakajima Y. Clinical significance and therapeutic potential of the programmed death-1 ligand/programmed death-1 pathway in human pancreatic cancer. Clin Cancer Res. 2007;13(7):2151–7. https://doi.org/10.1158/1078-0432.CCR-06-2746.
13. Ohigashi Y, Sho M, Yamada Y, Tsurui Y, Hamada K, Ikeda N, Mizuno T, Yoriki R, Kashizuka H, Yane K, Tsushima F, Otsuki N, Yagita H, Azuma M, Nakajima Y. Clinical significance of programmed death-1 ligand-1 and programmed death-1 ligand-2 expression in human esophageal cancer. Clin Cancer Res. 2005;11(8):2947–53. https://doi.org/10.1158/1078-0432.CCR-04-1469.
14. Iwai Y, Hamanishi J, Chamoto K, Honjo T. Cancer immunotherapies targeting the PD-1 signaling pathway. J Biomed Sci. 2017;24(1):26. https://doi.org/10.1186/s12929-017-0329-9.

15. Vanderstraeten A, Tuyaerts S, Amant F. The immune system in the normal endometrium and implications for endometrial cancer development. J Reprod Immunol. 2015;109:7–16. https://doi.org/10.1016/j.jri.2014.12.006.
16. Herzog TJ, Arguello D, Reddy SK, Gatalica G. PD-1, PD-L1 expression in 1599 gynecological cancers: implications for immunotherapy. Gynecol Oncol. 2015;137(suppl 1):204–5.
17. Brahmer JR, Tykodi SS, Chow LQ, Hwu WJ, Topalian SL, Hwu P, Drake CG, Camacho LH, Kauh J, Odunsi K, Pitot HC, Hamid O, Bhatia S, Martins R, Eaton K, Chen S, Salay TM, Alaparthy S, Grosso JF, Korman AJ, Parker SM, Agrawal S, Goldberg SM, Pardoll DM, Gupta A, Wigginton JM. Safety and activity of anti-PD-L1 antibody in patients with advanced cancer. N Engl J Med. 2012;366(26):2455–65. https://doi.org/10.1056/NEJMoa1200694.
18. Howitt BE, Shukla SA, Sholl LM, Ritterhouse LL, Watkins JC, Rodig S, Stover E, Strickland KC, D'Andrea AD, Wu CJ, Matulonis UA, Konstantinopoulos PA. Association of polymerase e-mutated and microsatellite-instable endometrial cancers with neoantigen load, number of tumor-infiltrating lymphocytes, and expression of PD-1 and PD-L1. JAMA Oncol. 2015;1(9):1319–23. https://doi.org/10.1001/jamaoncol.2015.2151.
19. Gargiulo P, Della Pepa C, Berardi S, Califano D, Scala S, Buonaguro L, Ciliberto G, Brauchli P, Pignata S. Tumor genotype and immune microenvironment in POLE-ultramutated and MSI-hypermutated endometrial cancers: new candidates for checkpoint blockade immunotherapy? Cancer Treat Rev. 2016;48:61–8. https://doi.org/10.1016/j.ctrv.2016.06.008.
20. Nelson BH, McAlpine JN. The more tumors change, the more they stay tame: do T cells keep POLE ultramutated endometrial carcinomas in check? Gynecol Oncol. 2015;138(1):1–2. https://doi.org/10.1016/j.ygyno.2015.06.004.
21. Bellone S, Centritto F, Black J, Schwab C, English D, Cocco E, Lopez S, Bonazzoli E, Predolini F, Ferrari F, Silasi DA, Ratner E, Azodi M, Schwartz PE, Santin AD. Polymerase ε (POLE) ultra-mutated tumors induce robust tumor-specific CD4+ T cell responses in endometrial cancer patients. Gynecol Oncol. 2015;138(1):11–7. https://doi.org/10.1016/j.ygyno.2015.04.027.
22. van Gool IC, Eggink FA, Freeman-Mills L, Stelloo E, Marchi E, de Bruyn M, Palles C, Nout RA, de Kroon CD, Osse EM, Klenerman P, Creutzberg CL, Tomlinson IP, Smit VT, Nijman HW, Bosse T, Church DN. POLE proofreading mutations elicit an antitumor immune response in endometrial cancer. Clin Cancer Res. 2015;21(14):3347–55. https://doi.org/10.1158/1078-0432.CCR-15-0057.
23. Eggink FA, Van Gool IC, Leary A, Pollock PM, Crosbie EJ, Mileshkin L, Jordanova ES, Adam J, Freeman-Mills L, Church DN, Creutzberg CL, De Bruyn M, Nijman HW, Bosse T. Immunological profiling of molecularly classified high-risk endometrial cancers identifies POLE-mutant and microsatellite unstable carcinomas as candidates for checkpoint inhibition. Onco Targets Ther. 2016;6(2):e1264565. https://doi.org/10.1080/2162402X.2016.1264565. eCollection 2017
24. de Jong RA, Leffers N, Boezen HM, ten Hoor KA, van der Zee AG, Hollema H, Nijman HW. Presence of tumor-infiltrating lymphocytes is an independent prognostic factor in type I and II endometrial cancer. Gynecol Oncol. 2009;114(1):105–10. https://doi.org/10.1016/j.ygyno.2009.03.022.
25. Cancer Genome Atlas Research Network, Kandoth C, Schultz N, Cherniack AD, Akbani R, Liu Y, Shen H, Robertson AG, Pashtan I, Shen R, Benz CC, Yau C, Laird PW, Ding L, Zhang W, Mills GB, Kucherlapati R, Mardis ER, Levine DA. Integrated genomic characterization of endometrial carcinoma. Nature. 2013;497(7447):67–73. https://doi.org/10.1038/nature12113.
26. Green AK, Feinberg J, Makker V. A review of immune checkpoint blockade therapy in endometrial cancer. Am Soc Clin Oncol Educ Book. 2020;40:1–7. https://doi.org/10.1200/EDBK_280503.
27. Ott PA, Bang YJ, Berton-Rigaud D, Elez E, Pishvaian MJ, Rugo HS, Puzanov I, Mehnert JM, Aung KL, Lopez J, Carrigan M, Saraf S, Chen M, Soria JC. Safety and antitumor activity of Pembrolizumab in advanced programmed death ligand 1-positive endometrial cancer: results

from the KEYNOTE-028 study. J Clin Oncol. 2017;35(22):2535–41. https://doi.org/10.1200/JCO.2017.72.5952.
28. Le DT, Uram JN, Wang H, Bartlett BR, Kemberling H, Eyring AD, Skora AD, Luber BS, Azad NS, Laheru D, Biedrzycki B, Donehower RC, Zaheer A, Fisher GA, Crocenzi TS, Lee JJ, Duffy SM, Goldberg RM, de la Chapelle A, Koshiji M, Bhaijee F, Huebner T, Hruban RH, Wood LD, Cuka N, Pardoll DM, Papadopoulos N, Kinzler KW, Zhou S, Cornish TC, Taube JM, Anders RA, Eshleman JR, Vogelstein B, Diaz LA Jr. PD-1 blockade in Tumors with mismatch-repair deficiency. N Engl J Med. 2015;372(26):2509–20. https://doi.org/10.1056/NEJMoa1500596.
29. Le DT, Durham JN, Smith KN, Wang H, Bartlett BR, Aulakh LK, Lu S, Kemberling H, Wilt C, Luber BS, Wong F, Azad NS, Rucki AA, Laheru D, Donehower R, Zaheer A, Fisher GA, Crocenzi TS, Lee JJ, Greten TF, Duffy AG, Ciombor KK, Eyring AD, Lam BH, Joe A, Kang SP, Holdhoff M, Danilova L, Cope L, Meyer C, Zhou S, Goldberg RM, Armstrong DK, Bever KM, Fader AN, Taube J, Housseau F, Spetzler D, Xiao N, Pardoll DM, Papadopoulos N, Kinzler KW, Eshleman JR, Vogelstein B, Anders RA, Diaz LA Jr. Mismatch repair deficiency predicts response of solid tumors to PD-1 blockade. Science. 2017;357(6349):409–13. https://doi.org/10.1126/science.aan6733.
30. Marabelle A, Le DT APA, Di Giacomo AM, De Jesus-Acosta A, Delord JP, Geva R, Gottfried M, Penel N, Hansen AR, Piha-Paul SA, Doi T, Gao B, Chung HC, Lopez-Martin J, Bang YJ, Frommer RS, Shah M, Ghori R, Joe AK, Pruitt SK, Diaz LA Jr. Efficacy of pembrolizumab in patients with noncolorectal high microsatellite instability/mismatch repair-deficient cancer: results from the phase II KEYNOTE-158 study. J Clin Oncol. 2020;38(1):1–10. https://doi.org/10.1200/JCO.19.02105.
31. Oaknin A, Ellard SL, Leath C III, Moreno V, Kristeleit R, Guo W, Lu S, Jenkins D, McEachern K, Jen Yu K, Dunlap S, Im E, Gilbert L. Preliminary safety, efficacy, and PK/PD characterization from GARNET, a phase 1 clinical trial of the anti-PD-1 monoclonal antibody, TSR-042, in patients with recurrent or advanced MSI-H endometrial cancer. Ann Oncol. 2018;29(suppl 8):viii334. https://doi.org/10.1093/annonc/mdy285.144.
32. Oaknin A, Pothuri B, Gilbert L, Sabatier R, Brown J, Ghamande S, Mathews C, O'Malley DM, Kristeleit R, Boni V, Gravina A, Banerjee S, Miller R, Pikiel J, Mirza MR, Dewal N, Antony G, Dong Y, Zografos E, Veneris J, Tinker AV. Safety, efficacy, and biomarker analyses of dostarlimab in patients with endometrial cancer: interim results of the phase I GARNET study. Clin Cancer Res. 2023;29(22):4564–74. https://doi.org/10.1158/1078-0432.CCR-22-3915.
33. Antill Y, Kok PS, Robledo K, Yip S, Cummins M, Smith D, Spurdle A, Barnes E, Lee YC, Friedlander M, Baron-Hay S, Shannon C, Coward J, Beale P, Goss G, Meniawy T, Lombard J, Andrews J, Stockler MR, Mileshkin L, Australia New Zealand Gynaecological Oncology Group (ANZGOG). Clinical activity of durvalumab for patients with advanced mismatch repair-deficient and repair-proficient endometrial cancer. A nonrandomized phase 2 clinical trial. J Immunother Cancer. 2021;9(6):e002255. https://doi.org/10.1136/jitc-2020-002255.
34. Konstantinopoulos PA, Luo W, Liu JF, Gulhan DC, Krasner C, Ishizuka JJ, Gockley AA, Buss M, Growdon WB, Crowe H, Campos S, Lindeman NI, Hill S, Stover E, Schumer S, Wright AA, Curtis J, Quinn R, Whalen C, Gray KP, Penson RT, Cannistra SA, Fleming GF, Matulonis UA. Phase II study of avelumab in patients with mismatch repair deficient and mismatch repair proficient recurrent/persistent endometrial cancer. J Clin Oncol. 2019;37(30):2786–94. https://doi.org/10.1200/JCO.19.01021.
35. Konstantinopoulos PA, Gockley AA, Xiong N, Krasner C, Horowitz N, Campos S, Wright AA, Liu JF, Shea M, Yeku O, Castro C, Polak M, Lee EK, Sawyer H, Bowes B, Moroney J, Cheng SC, Tayob N, Bouberhan S, Spriggs D, Penson RT, Fleming GF, Nucci MR, Matulonis UA. Evaluation of treatment with talazoparib and avelumab in patients with recurrent mismatch repair proficient endometrial cancer. JAMA Oncol. 2022;8(9):1317–22. https://doi.org/10.1001/jamaoncol.2022.2181.
36. Makker V, Colombo N, Casado Herráez A, Monk BJ, Mackay H, Santin AD, Miller DS, Moore RG, Baron-Hay S, Ray-Coquard I, Ushijima K, Yonemori K, Kim YM, Guerra Alia EM, Sanli

UA, Bird S, Orlowski R, McKenzie J, Okpara C, Barresi G, Lorusso D. Lenvatinib plus pembrolizumab in previously treated advanced endometrial cancer: updated efficacy and safety from the randomized phase III study 309/KEYNOTE-775. J Clin Oncol. 2023;41(16):2904–10. https://doi.org/10.1200/JCO.22.02152.
37. Eskander RN, Sill MW, Beffa L, Moore RG, Hope JM, Musa FB, Mannel R, Shahin MS, Cantuaria GH, Girda E, Mathews C, Kavecansky J, Leath CA 3rd, Gien LT, Hinchcliff EM, Lele SB, Landrum LM, Backes F, O'Cearbhaill RE, Al Baghdadi T, Hill EK, Thaker PH, John VS, Welch S, Fader AN, Powell MA, Aghajanian C. Pembrolizumab plus chemotherapy in advanced endometrial cancer. N Engl J Med. 2023;388(23):2159–70. https://doi.org/10.1056/NEJMoa2302312.
38. Powell MA, Bjørge L, Willmott L, Novák Z, Black D, Gilbert L, Sharma S, Valabrega G, Landrum LM, Gropp-Meier M, Stuckey A, Boere I, Gold MA, Segev Y, Gill SE, Gennigens C, Sebastianelli A, Shahin MS, Pothuri B, Monk BJ, Buscema J, Coleman RL, Slomovitz BM, Ring KL, Herzog TJ, Balas MM, Grimshaw M, Stevens S, Lai DW, McCourt C, Mirza MR. Overall survival in patients with endometrial cancer treated with dostarlimab plus carboplatin-paclitaxel in the randomized ENGOT-EN6/GOG-3031/RUBY trial. Ann Oncol. 2024;35(8):728–38. https://doi.org/10.1016/j.annonc.2024.05.546.
39. Colombo N, Biagioli E, Harano K, Galli F, Hudson E, Antill Y, Choi CH, Rabaglio M, Marmé F, Marth C, Parma G, Fariñas-Madrid L, Nishio S, Allan K, Lee YC, Piovano E, Pardo B, Nakagawa S, McQueen J, Zamagni C, Manso L, Takehara K, Tasca G, Ferrero A, Tognon G, Lissoni AA, Petrella M, Laudani ME, Rulli E, Uggeri S, Barretina Ginesta MP, AtTEnd study group. Atezolizumab and chemotherapy for advanced or recurrent endometrial cancer (AtTEnd): a randomised, double-blind, placebo-controlled, phase 3 trial. Lancet Oncol. 2024;25(9):1135–46. https://doi.org/10.1016/S1470-2045(24)00334-6.
40. Marth C, Moore RG, Bidziński M, Pignata S, Ayhan A, Rubio MJ, Beiner M, Hall M, Vulsteke C, Braicu EI, Sonoda K, Wu X, Frentzas S, Mattar A, Lheureux S, Chen X, Hasegawa K, Magallanes-Maciel M, Choi CH, Shalkova M, Kaen D, Wang PH, Berger R, Okpara CE, McKenzie J, Yao L, Orlowski R, Khemka V, Gilbert L, Makker V, ENGOT-en9/LEAP-001 Investigators. First-line lenvatinib plus pembrolizumab versus chemotherapy for advanced endometrial cancer: a randomized, open-label, phase III trial. J Clin Oncol. 2024:JCO2401326. https://doi.org/10.1200/JCO-24-01326.
41. Westin SN, Moore K, Chon HS, Lee JY, Thomes Pepin J, Sundborg M, Shai A, de la Garza J, Nishio S, Gold MA, Wang K, McIntyre K, Tillmanns TD, Blank SV, Liu JH, McCollum M, Contreras Mejia F, Nishikawa T, Pennington K, Novak Z, De Melo AC, Sehouli J, Klasa-Mazurkiewicz D, Papadimitriou C, Gil-Martin M, Brasiuniene B, Donnelly C, Del Rosario PM, Liu X, Van Nieuwenhuysen E, DUO-E Investigators. Durvalumab plus carboplatin/paclitaxel followed by maintenance durvalumab with or without olaparib as first-line treatment for advanced endometrial cancer: the phase III DUO-E trial. J Clin Oncol. 2024;42(3):283–99. https://doi.org/10.1200/JCO.23.02132.
42. Mirza MR, Ghamande S, Hanker LC, Black D, Raaschou-Jensen N, Gilbert L, Oaknin A, Secord AA, Savarese A, Holloway RW, Kristeleit1 R, Buscema J, Boere LA, Sharma S, et al. Progression-free survival (PFS) in primary advanced or recurrent endometrial cancer (pA/rEC) in the overall and mismatch repair proficient (MMR/MSS) populations and in histological and molecular subgroups: results from part 2 of the RUBY trial. ESMO Open. 2024;9(Suppl 5103538) June2024 ESMO Gynecologic Cancers Congress. https://www.esmoopen.com/article/S2059-7029(24)01307-3/fulltext
43. Slomovitz BM, Cibula D, Simsek T, Mirza MR, Maćkowiak-Matejczk B, Hudson E, Romero I, Colombo N, Korach J, Yin R, Gilbert L, Hasegawa K, Tyulyandina A, Baron-Hay SE, Willmott L, Backes FJ, et al. J Clin Oncol. 2022;40(16 Suppl) https://doi.org/10.1200/JCO.2022.40.16_suppl.TPS562.
44. Rubio MJ, Paoletti X, Davis AJ, Hudson E, Lorusso D, Tognon G, Hasler-Strub U, Choi CH, Vardar MA, Lheureux S, Gorp TV, DSP T, Trillsch F, et al. J Clin Oncol. 2023;41(16 Suppl) https://doi.org/10.1200/JCO.2023.41.16_suppl.TPS5630.

45. RAINBO Research Consortium. Refining adjuvant treatment in endometrial cancer based on molecular features: the RAINBO clinical trial program. Int J Gynecol Cancer. 2023;33(1):109–17. https://doi.org/10.1136/ijgc-2022-004039.
46. Atkinson VG, Antras JF, Bedard P, Brenner W, Brown J, Lemech CR, Lloyd P, Margolin K, Mathew M, Park JJ, Thomas S, Twardowski P, Gardner H, Prawira A, Coello M, Ebrahimimizadeh W, To MD, Merchant R, Karanam SM, Yavari A, Siu LL, Tawbi HA, Ascierto PA. Interim results from the phase 1/2 ABILITY-1 study of a long-acting 'beta-enhanced not-alpha' IL-2 superkine in patients with advanced solid tumors Cancer Res (2025) 85 (8_Supplement_2): CT047. https://doi.org/10.1158/1538-7445.AM2025-CT047.
47. Cantillo E, Zemla TJ, Moroney JW, Alvarez EA, Shah J, Lin YG, DiLallo G, Liu JF, Mahdi H, Secord AA, Nixon AB, Mandrekar SJ, Matulonis UA, Kozono DE, Slomovitz BM. AFT-50 EndoMAP: a phase IB/II multicohort study of targeted agents and/or immunotherapy for patients with recurrent or persistent endometrial cancer (EC). J Clin Oncol. 2024;42(16 Suppl). https://ascopubs.org/doi/10.1200/JCO.2024.42.16_suppl.TPS5642.
48. Shum E, Reilley M, Najjar Y, et al. 523 Preliminary clinical experience with XmAb20717, a PD-1 x CTLA-4 bispecific antibody, in patients with advanced solid tumors. Journal for ImmunoTherapy of Cancer. 2021;9: https://doi.org/10.1136/jitc2021-SITC2021.523.
49. Kelley RK, Van Cutsem E, Lee MS, Wolf I, Fakih M, de Vos-Geelen J, Lee V, Vogel A, Wu XL, Jin F, Naik GS, O'Reilly EM. Phase 2 open-label study of pembrolizumab plus lenvatinib and belzutifan in patients with advanced solid tumors. J Clin Oncol. 2022;40(4 Suppl). https://doi.org/10.1200/JCO.2022.40.4_suppl.TPS669.
50. Ayers M, Lunceford J, Nebozhyn M, Murphy E, Loboda A, Kaufman DR, Albright A, Cheng JD, Kang SP, Shankaran V, Piha-Paul SA, Yearley J, Seiwert TY, Ribas A, TK MC. IFN-γ-related mRNA profile predicts clinical response to PD-1 blockade. J Clin Invest. 2017;127(8):2930–40. https://doi.org/10.1172/JCI91190.
51. Alley EW, Lopez J, Santoro A, Morosky A, Saraf S, Piperdi B, van Brummelen E. Clinical safety and activity of pembrolizumab in patients with malignant pleural mesothelioma (KEYNOTE-028): preliminary results from a non-randomised, open-label, phase 1b trial. Lancet Oncol. 2017;18(5):623–30. https://doi.org/10.1016/S1470-2045(17)30169-9.
52. Cristescu R, Nebozhyn M, Zhang C, Albright A, Kobie J, Huang L, Zhao Q, Wang A, Ma H, Alexander Cao Z, Morrissey M, Ribas A, Grivas P, Cescon DW, TK MC, Snyder A, Ayers M, Lunceford J, Loboda A. Transcriptomic determinants of response to pembrolizumab monotherapy across solid tumor types. Clin Cancer Res. 2022;28(8):1680–9. https://doi.org/10.1158/1078-0432.CCR-21-3329.
53. Makker V, Taylor MH, Aghajanian C, Cohn AL, Brose MS, Simone CD, Cao ZA, Suttner L, Loboda A, Cristescu R, Jelinic P, Orlowski R, Dutta L, Matsui J, Dutcus CE, Minoshima Y, Messing MJ. Evaluation of potential biomarkers for lenvatinib plus pembrolizumab among patients with advanced endometrial cancer: results from study 111/KEYNOTE-146. J Immunother Cancer. 2024;12(1):e007929. https://doi.org/10.1136/jitc-2023-007929.
54. Westin S, Moore K, Chon HS, Pepin JT, Salinas E, Starks D, Disilvestro P, Slomovitz B, Vettus E, Montosa FG, Agyemang-Prempeh K, Zagouri F, Kim JW, Gao Q, et al. Durvalumab plus carboplatin/paclitaxel followed by durvalumab with/without olaparib in endometrial cancer: exploratory analyses of biomarker/histological heterogeneity and efficacy in the DUO-E mismatch repair proficient subpopulation. Int J Gynecol Cancer. 2024;34(Suppl. 3):A4–A5LB002/#1594. https://www.sciencedirect.com/science/article/abs/pii/S1048891X24162342
55. Mirza MR, Sharma S, Herrstedt J, Shahin MS, Cibula D, Fleming E, Raspagliesi F, Buscema J, Hanker LC, Coleman RL, Boerel RL, Schneider K, et al. Dostarlimab + chemotherapy for the treatment of primary advanced or recurrent endometrial cancer: analysis of progression-free survival and overall survival outcomes by molecular classification in the ENGOT-EN6-NSGO/GOG-3031/RUBY trial. Ann Oncol. 2023;34(Suppl 2):S507–42. https://doi.org/10.1016/S0923-7534(23)01937-3. https://www.targetedonc.com/view/three-of-4-molecular-subgroups-respond-to-dostarlimab-in-advanced-endometrial-carcinoma

56. Yamamura A, Hamanishi J, Yamanoi K, Sunada M, Taki M, Mizuno R, Okada Y, Murakami R, Aisu Y, Maekawa H, Yamaguchi K, Mandai M. Colorectal anastomotic leakage after conversion surgery for advanced endometrial cancer treated with lenvatinib plus pembrolizumab: a case report. Int Cancer Conf J. 2024;14(1):64–71. https://doi.org/10.1007/s13691-024-00739-6. eCollection 2025 Jan
57. Ohira G, Endo S, Imanishi S, Tochigi T, Maruyama T, Hayano K, Maruyama M, Matsubara H. Prognosis and predictive factors of conversion surgery for initially unresectable advanced colorectal cancer. Langenbeck's Arch Surg. 2024;409(1):182. https://doi.org/10.1007/s00423-024-03374-0.
58. Kim TH, Uyama I, Rha SY, Bencivenga M, An J, Wyrwicz L, Koo DH, van Hillegersberg R, Lee KW, Li G, Yoshikawa T, Badgwell B, Lorenzen S, Kim IH, Lee IS, Han HS, Hoon H. Conversion therapy for stage IV gastric cancer: report from the expert consensus meeting at KINGCA WEEK 2024. J Gastric Cancer. 2025;25(1):133–52. https://doi.org/10.5230/jgc.2025.25.e9.
59. https://rctportal.mhlw.go.jp/en/detail?trial_id=UMIN000055120.

Novel Treatment for Endometrial Cancer

10

Takashi Mitamura

Abstract

The mainstay of advances in novel treatments for endometrial cancer is immune checkpoint inhibitors (ICIs). However, the objective response rate of ICI-related treatments is approximately 40% for endometrial cancer, and patients still require novel therapies. Ongoing and other latest phase II/III clinical trials include a selective MDM2 inhibitor; a selective inhibitor of XPO1; mTOR inhibitors; a monoclonal antibody directed against HER2; a selective TIE2 inhibitor; antibody-drug conjugates (ADCs) targeting TROP2, FRα, HER2, HER3, and TF; and an alkylating histone deacetylase inhibitor (HDACi). ADCs are expected to become primary candidates for new treatments for endometrial cancer. In addition, it is also noteworthy that progress has been made in both making efficient use of existing antiangiogenic agents and in developing new drugs targeting p53, which has been a long-standing goal since previous studies have indicated that p53 mutations occur in endometrial cancer. In the future, we also need to verify the introduction and application of cancer genome profiling using whole-genome sequencing and AI systems for treatment individualization.

Keywords

Clinical trial · Antibody-drug conjugates · Molecular-targeted therapy

---

T. Mitamura (✉)
Department of Obstetrics and Gynecology, Hokkaido University Graduate School of Medicine, Sapporo, Japan
e-mail: takami@huhp.hokudai.ac.jp

H. Watari, T. Mitamura (eds.), *Uterine Body Neoplasms*, Comprehensive Gynecology and Obstetrics, https://doi.org/10.1007/978-981-95-0946-1_10

10.1 Introduction

Surgery is the first-choice treatment for endometrial cancer. After surgery, adjuvant therapy is selected based on the assessment of recurrence risk. Risk factors include the stage of progression at surgery, histological type, histological grade, pelvic and para-aortic lymph node metastasis, muscle invasion, lymphovascular invasion, cervical stromal invasion, adnexal, serosa, parametrium invasion, vaginal wall invasion, bladder and rectal invasion, intraperitoneal dissemination, and distant metastasis. Adjuvant therapy is indicated in order to achieve long-term survival based on the abovementioned risk factors. In the treatment of advanced cases that are expected to be inoperable, an individualized treatment plan using surgery, chemotherapy, radiotherapy, or hormonal therapy is required, depending on the individual situation. As for chemotherapy, combinations of doxorubicin (Adriamycin) + cisplatin and of paclitaxel and carboplatin have been considered the standard treatment based on the results of GOG107 [1] and GOG209 [2]. Drug therapy is also used in cases of multiple or localized recurrence, in which surgery or radiotherapy is not an option because of previous treatment or the site of recurrence. However, no treatments have been shown to sufficiently improve the survival once recurrence occurs—at least not until immune checkpoint inhibitors (ICIs) were developed. The mainstay of advances in novel treatments for endometrial cancer is drug therapy, particularly ICIs. These agents improve the survival of patients with advanced and recurrent endometrial cancer, for which only limited modalities, such as chemotherapy and radiotherapy, are available. However, the objective response rate of ICI-related treatments is approximately 40% (see Chap. 9) for endometrial cancer, and patients still require novel therapies.

In 2024, the crude numbers of ongoing clinical trials for endometrial cancer were 80, 59, 114, 15, and 34 in phases I, I/II, II, II/III, and III, respectively. Among them, the crude proportion was highest for molecular-targeted therapy, which included ICIs (65.6%), followed by antibody-drug conjugates (15.2%), hormonal therapy (11.9%), radiotherapy (4.6%), immune therapy other than ICIs (1.3%), and others (1.3%). This chapter introduces the latest trends in the development of treatments for endometrial cancer, other than ICIs.

10.2 Molecular-Targeted Therapy

Cytotoxic agents remain the primary choice for endometrial cancer treatment. However, the approval of novel cytotoxic agents has slowed in the past decades because of their limited ability to improve survival outcomes, especially in recurrent lesions. Paradigm shifts in cancer treatment have accelerated the individualization of targeted therapy based on genetic alterations that characterize carcinogenesis, aberrant growth, and drug resistance. As for endometrial cancer, there are few drugs that inhibit several targets, such as epidermal growth factor receptor (EGFR), platelet-derived growth factor receptor (PDGFR), BRAF, HER2, the mTOR/AKT/

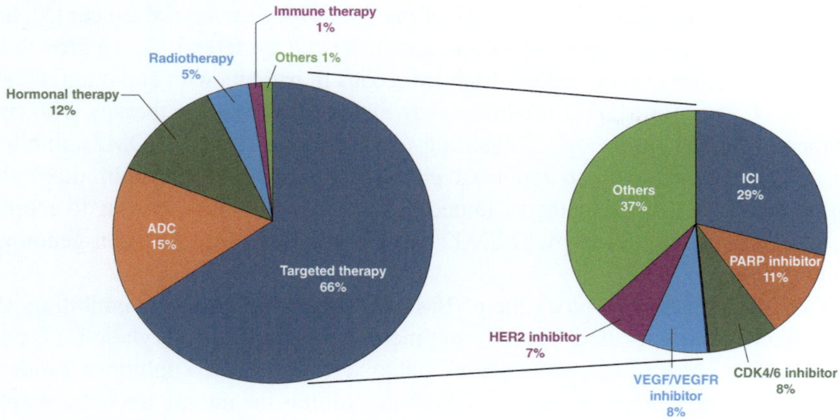

Fig. 10.1 The direction of establishment of novel treatments for endometrial cancer by estimating with the ongoing phase II/III clinical trial

PI3K pathway, or ALK, for which effective targeted therapies are available in other cancers and show a high objective response as a single agent in phase III clinical trials. Recent clinical trials have estimated that molecular-targeted therapy accounts for over 60% of new treatments for endometrial cancer (Fig. 10.1). The mainstay of treatment is ICIs and PARP inhibitors combined with other drugs, followed by CDK4/6 inhibitors. This section introduces several unique drugs used in ongoing or completed clinical trials.

10.2.1 Phase III Clinical Trials

10.2.1.1 Navtemadlin (AMG-232)

The p53 signaling pathway is a critical mediator of cell fate following genotoxic damage and provides key tumor suppressor functions. Under cellular stress and DNA damage, p53 functions as a transcriptional activator and binds to the promoters of hundreds of genes, including those crucial for genotoxic stress recovery, senescence initiation, and apoptosis activation. In contrast, the *TP53* gene, encoding p53, is the most frequently mutated gene in all types of cancer. Tumorigenic gain-of-function mutations have also been reported in *TP53*. A previous study showed that approximately 70% of endometrial cancers had wild-type *TP53* [3]. As high p53 levels can be detrimental to the growth and development of normal cells, intracellular p53 levels are tightly regulated. The primary suppressor MDM2 prevents p53 from regulating the transcription of downstream target genes by binding to p53 directly, promotes SUMOylation and nuclear export of p53 by interacting with various cytoplasmic proteins, and promotes ubiquitination for proteasomal degradation as an E3 ubiquitin ligase that targets p53 [4].

MDM2 is upregulated in up to 50% of patients with endometrial cancer [5], and augmenting the original p53 activity to arrest the cell cycle and reduce growth by MDM2 inhibition is expected to be advantageous in patients with endometrial cancer. The strategy for MDM2 inhibitors may include combination therapy with chemotherapy and monotherapy. Regarding combination therapy, MDM2 inhibitors can sensitize cancer cells to genotoxic chemotherapy, such as cisplatin, doxorubicin, and etoposide, by inducing sustained damage without repair, leading to apoptosis. Regarding monotherapy, MDM2 inhibitors may aim for non-genotoxic support of p53.

Two mechanisms can block the p53/MDM2 interaction: first, by inhibiting p53 and MDM2 binding, and second, by increasing the phosphorylation of p53. Navtemadlin (AMG-232), an orally available selective MDM2 inhibitor, binds to the hydrophobic binding pocket of MDM2 and inhibits the interaction between p53 and MDM2, resulting in the increased stability and transcriptional activity of p53. ENGOT-EN21/GOG-3089 is an ongoing global two-part phase II/III study evaluating the safety and efficacy of Navtemadlin as maintenance therapy in *TP53* wild-type advanced and recurrent EC patients who responded to chemotherapy. The strategy of this trial was to inhibit the recurrence of non-genotoxic p53 supported by Navtemadlin after genotoxic chemotherapy. Eligible patients completed chemotherapy and achieved a complete or partial response, and they received Navtemadlin in their maintenance therapy (NCT05797831). Regarding safety, the dose-finding phase Ib/II KRT-232-103 study [6] in patients with Merkel cell carcinoma demonstrated treatment-emergent adverse events (TEAEs) in 100% (68% grade 3/4) of patients. The most common grade 3/4 TEAEs were hematologic (32% anemia, 32% lymphopenia, and 19% thrombocytopenia).

10.2.1.2 Anlotinib

Anlotinib (Catequentinib) is a receptor tyrosine kinase (RTK) inhibitor with potential antineoplastic and antiangiogenic activities that target vascular endothelial growth factor receptor (VEGFR)1, VEGFR2, VEGFR3, PDGFR-α, and fibroblast growth factor receptors (FGFR1, FGFR2, and FGFR3) [7]. These broad targeting mechanisms may be more advantageous than those of other conventional RTK inhibitors [8]. Furthermore, anlotinib showed strong inhibitory activity against the stem cell factor receptor c-Kit, suppressing tumor growth. Therefore, anlotinib can inhibit tumor angiogenesis in multiple ways, and it achieves tumor suppression via multiple pathways while avoiding possible target resistance that may lead to treatment failure [9]. In terms of adverse effects, the ALTER 0303 phase III randomized clinical trial demonstrated that adverse events of grade ≥ 3 were reported in 61.9% of patients in the anlotinib group, with the most common adverse events being hypertension (13.6%), hyponatremia (8.2%), and elevated γ-glutamyltransferase (5.4%) [10].

Anlotinib is approved in China for the treatment of patients with locally advanced or metastatic non-small-cell lung cancer (NSCLC) who have undergone progression or recurrence after two or more lines of systemic chemotherapy [11]. In a phase II trial of 56 patients with recurrent or metastatic endometrial cancer, anlotinib showed an overall objective response rate of 42.9% [12]. Although we cannot directly compare the benefit with other antiangiogenic agents, including bevacizumab and lenvatinib, for which the reported objective responses were 13.5% [13] and 14.3% (NCT01111461) for endometrial cancer, anlotinib may have a high potential to repress advanced endometrial cancer.

A phase III trial is ongoing in patients with recurrent or metastatic endometrial cancer to evaluate the efficacy and safety of anlotinib plus chemotherapy (weekly paclitaxel, pegylated liposomal doxorubicin, or topotecan) versus chemotherapy alone (NCT02584478). Another strategy is combination therapy with anlotinib and ICIs. It is suggested that the VEGF-VEGFR pathway, which is targeted by anlotinib, suppresses anti-cancer immunity by inducing suppressor T cells and myeloid-derived suppressor cells and changing M1/M2 macrophage polarization [14]. Therefore, the concurrent use of anti-VEGFR drugs and ICIs makes sense, and the combination therapy of antiangiogenic lenvatinib, which targets VEGFR, FGFR, and PDGFR, with pembrolizumab, which targets PD-1, has demonstrated a significant impact in recurrent endometrial cancer [15]. Ongoing clinical trials have investigated the significance of anlotinib with an anti-PD-L1 antibody in patients with relapsed platinum-resistant ovarian cancer, advanced renal cancer, and standard-chemo-immunotherapy-failed lung cancer (NCT05145218, NCT04523272, NCT04322617).

10.2.1.3 Selinexor (Box in Fig. 10.2)

Selinexor is a selective inhibitor of XPO1-mediated nuclear export, which is responsible for the export of proteins, including tumor suppressor proteins, such as p53, PTEN, and FOXO1, along with growth/cell cycle regulatory proteins, such as p21 and p27 [16]. Selinexor is used in combination with dexamethasone for the treatment of adult patients with relapsed or refractory multiple myeloma and bortezomib and dexamethasone for the treatment of adult patients with multiple myeloma who have received at least one prior therapy.

A previous study suggested that endometrial cancer overexpressed XPO1, which is possibly related to copy number amplification. Therefore, inhibition of XPO1 by selinexor theoretically leads to nuclear accumulation and functional reactivation of tumor suppressor proteins, such as p53, in endometrial cancer [16]. A previous study also showed that endometrial cancer with wild-type *TP53* comprises approximately 70% of all patients [3] and suggested the existence of many patients with indications for selinexor. A randomized, prospective, multicenter, double-blinded, placebo-controlled, phase III study, ENGOT-EN5/GOG-3055/SIENDO showed that selinexor had a significant modest effect on improving the progression-free survival (PFS), and the median PFS was 5.7 (95% confidence interval [CI] 3.81 to 9.20) months with selinexor vs. 3.8 (95% CI 3.68 to 7.39) months with placebo (hazard ratio [HR] 0.76). Interestingly, the patients with the wild-type *TP53* endometrial cancer showed a higher impact than those with *TP53* mutant or aberrant and had a median PFS of 13.7 and 3.7 months with selinexor and placebo (HR 0.41), respectively. The exploratory subgroup analysis of long-term follow-up in the patients with wild-type *TP53* endometrial cancer demonstrated that a benefit in the

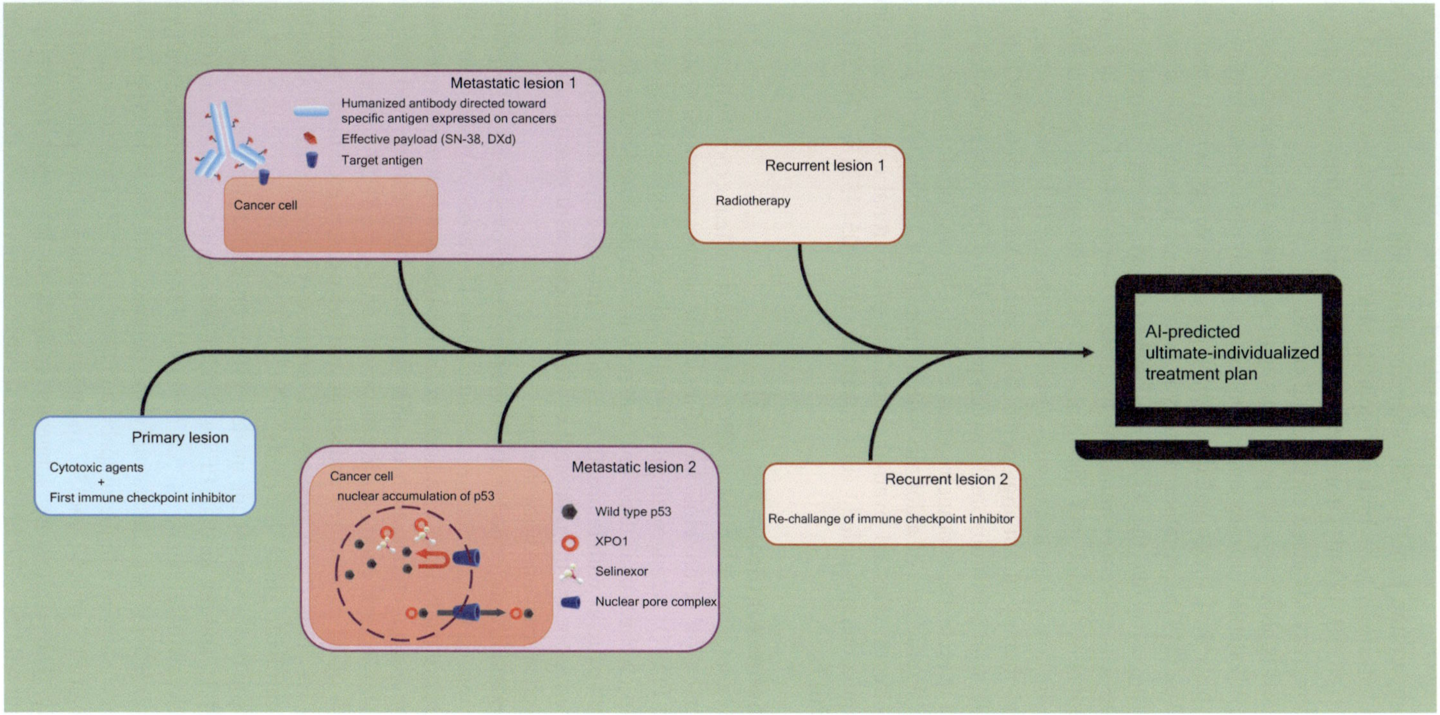

Fig. 10.2 Representative novel treatments for endometrial cancer and ultimate individualization using artificial intelligence

PFS was seen with the selinexor group vs. placebo, regardless of the MMR status (MMR proficient: 39.5 vs. 4.9 months, HR 0.36; 95% CI 0.19–0.71; MMR deficient: 13.1 vs. 3.7 months, HR 0.49; 95% CI 0.18–1.34) [17].

Because most ongoing phase III clinical trials of selinexor are conducted in myeloproliferative neoplasms, leukemia, and lymphoma, it is noteworthy that solid cancers are sensitive to selinexor. The most common grade 3 treatment-related adverse events were nausea (9%), neutropenia (9%), and thrombocytopenia (7%). There was also one case of treatment-related grade 4 thrombocytopenia with no severe bleeding and no cases of febrile neutropenia [17].

10.2.2 Phase II Clinical Trials with Promising Results

10.2.2.1 mTOR Inhibitor

activityABTL0812 (ibrilatazar) is an orally bioavailable α-hydroxylinoleic acid derivative that activates PPARα/γ and upregulates TRIB3, thereby functionally inhibiting the PI3K–Akt–mTOR pathway and inducing cytotoxic autophagy. The phase I/II ENDOLUNG trial aimed to determine the safety and efficacy of ibrilatazar combined with paclitaxel and carboplatin in patients with advanced/recurrent endometrial cancer. The combination of ABTL0812 plus paclitaxel and carboplatin showed an overall response rate of 65.8% (13.2% complete response and 52.6% partial response), with a median duration of response of 7.4 months. The most frequent hematological adverse events were neutropenia (52.9%), anemia (37.3%), and thrombocytopenia (19.6%). Nausea (66.7%), asthenia (66.7%), diarrhea (54.9%), and vomiting (54.9%) were the most frequent non-hematological adverse events [18].

Everolimus is a natural macrocyclic lactone sirolimus derivative with immunosuppressant and antiangiogenic properties. It binds to immunophilin FK binding protein-12 (FKBP-12) to generate an immunosuppressive complex that binds to and inhibits the activation of mTOR. Everolimus is indicated for the treatment of patients with advanced hormone receptor-positive human epidermal growth factor receptor 2 (HER2)-negative breast cancer, progressive neuroendocrine tumors of pancreatic origin (PNET), progressive neuroendocrine tumors (NET) of gastrointestinal or lung origin, advanced renal cell carcinoma, and tuberous sclerosis. A single-stage, open-label, two-arm randomized phase II trial of women with advanced or persistent or recurrent endometrial cancer showed that the combination of everolimus and letrozole had a 22% response rate, suggesting the significance of the combination of mTOR inhibitors and hormonal therapy [19].

In addition, combination therapy with the mTOR inhibitor vistusertib with anastrozole, which selectively binds to and reversibly inhibits aromatase to inhibit estrogen synthesis, showed an overall response rate of 24.5% and a median PFS of 5.2 months in patients with hormone receptor-positive recurrent or metastatic endometrial cancer. Fatigue, lymphopenia, hyperglycemia, and diarrhea were the most common adverse events associated with a vistusertib grade of ≥2 [20].

In contrast, another phase II study to investigate the clinical efficacy of sapanisertib (a selective dual mTORC1/2 inhibitor) in combination with paclitaxel in advanced, recurrent, or persistent endometrial cancer, demonstrated that the median PFS was 3.7 (paclitaxel) vs. 5.6 months (paclitaxel + sapanisertib) after a median follow-up of 14.4 vs. 17.2 months, respectively, and the primary endpoint was not met [21].

10.2.2.2 Trastuzumab

Trastuzumab is a monoclonal antibody directed against HER2, which is overexpressed in approximately 30% of uterine serous carcinomas [22]. A randomized phase II trial of carboplatin-paclitaxel (C/P) compared to carboplatin-paclitaxel-trastuzumab (C/P/T) in stage III–IV or recurrent uterine serous carcinomas that overexpress HER2/neu demonstrated that the overall survival (OS) was higher in the C/P/T arm than in the C/P arm, with a median of 29.6 vs. 24.4 months (HR 0.58). The benefit was most notable in those with stage III–IV disease, with a median survival not reached in the C/P/T arm vs. 24.4 months in the C/P arm (HR 0.49). Addition of T to C/P increased the PFS and OS in women with advanced/recurrent HER2/neu-positive uterine serous carcinoma, with the greatest benefit observed in the treatment of stage III–IV disease. Toxicity did not differ markedly between the arms, and no patient required discontinuation of trastuzumab for toxicity. Based on these results, the addition of trastuzumab to cytotoxic agents may represent a new standard treatment for uterine serous carcinoma tumors that overexpress HER2/neu, particularly in women with advanced primary disease [23].

10.2.2.3 Rebastinib

Rebastinib is a first-in-class investigational, orally administered, selective switch-control kinase inhibitor of tunica interna endothelial cell kinase (TIE2). An open-label, multicenter phase Ib/II study of rebastinib in combination with paclitaxel in patients with advanced or metastatic endometrial cancer showed that the overall response rate was 33%, and serious adverse events (SAEs), at least those possibly related to rebastinib, included muscular weakness, nausea, acute myocardial infarction, atrial flutter, dehydration, non-infective encephalitis, peritonitis, and stress cardiomyopathy, which resolved after dose interruption [24].

10.2.2.4 Bevacizumab

Single-agent bevacizumab is not significantly effective in patients with endometrial cancer, as mentioned above. Combination therapy with other agents did not show any significant effects. In the MITO END-2 study [25], patients with FIGO stage III–IV or recurrent endometrial cancer were randomized to receive a carboplatin-paclitaxel standard dose for 6–8 cycles or carboplatin-paclitaxel and bevacizumab in combination with chemotherapy and maintenance until disease progression or unacceptable toxicity. The PFS (10.5 vs. 13.7 months), overall response rate (53.1% vs. 74.4%), and OS (29.7 vs. 40.0 months) showed non-significant increases in bevacizumab-treated patients. The GOG-86P study (NCT00977574) was a three-arm, randomized phase II study of paclitaxel/carboplatin/bevacizumab, paclitaxel/

carboplatin/temsirolimus, or ixabepilone/carboplatin/bevacizumab as initial therapy for advanced or recurrent endometrial cancer; there was also no statistically significant survival benefit of adding bevacizumab to chemotherapy overall [26]. However, an additional study of GOG-86P showed that patients whose tumors harbored mutant *TP53*, as confirmed by p53 overexpression on immunohistochemistry, benefited significantly from bevacizumab when added to chemotherapy in the upfront setting, with an HR of 0.31 for the OS [27]. This new knowledge is important for patients with p53-mutated tumors who need better individualized treatment, as their prognosis is worse than that of patients with wild-type *TP53*.

10.3 Antibody-Drug Conjugate (ADC) (Box in Fig. 10.2)

Data from ongoing clinical trials also suggest that the interest in drug development for endometrial cancer is shifting to ADCs. ADCs are a targeted cancer treatment that combine a monoclonal antibody with a chemotherapy drug to deliver the drug directly to tumor cells with high selectivity, and they frequently show a higher efficacy than conventional cytotoxic agents. The development of ADCs tested in the ongoing phase 3 clinical trial for endometrial cancer began in the early 2010s (Fig. 10.3). Disitamab vedotin, an ADC comprising a monoclonal antibody against HER2 conjugated via a cleavable linker to the cytotoxic agent monomethyl auristatin E, has been approved in China for the treatment of gastric and urothelial cancers since 2021 [28].

In endometrial cancer, targeted tumor antigens have recently narrowed down to tumor-associated calcium signal transducer 2 (TROP2), folate receptor alpha (FRα), erb-b2 receptor tyrosine kinase 2 (HER2), erb-b2 receptor tyrosine kinase 3 (HER3), and tissue factor (TF) (Fig. 10.4). TROP2 is a transmembrane protein that is overexpressed in various tumors. Its expression is associated with enhanced tumor aggressiveness, metastasis, drug resistance, and an increased tumor cell survival. The expression of TROP2 was found in up to 92% of endometrial cancer cases [29–32]. One previous study demonstrated that the expression of TROP2

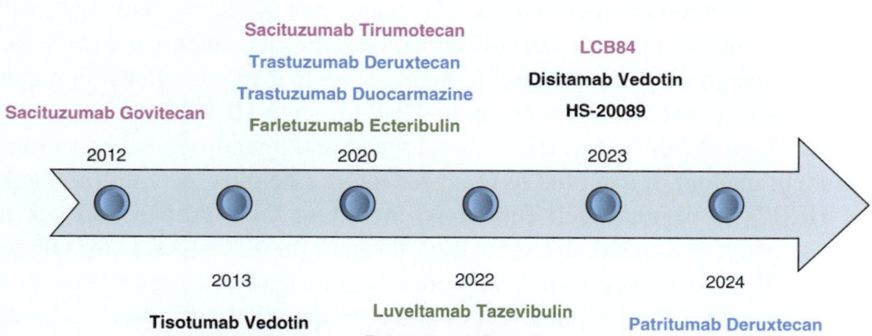

Fig. 10.3 Development of antibody-drug conjugates tested in a phase III clinical trial for endometrial cancer

Target antigen	Expression in endometrial cancer
TROP2 (Sacituzumab, Hu2G10, Datopotamab)	- 92.0 %
FRα (Farletuzumab, Luveltamab)	50.5 - 89.0 %
HER2 (Trastuzumab Deruxtecan, Trastuzumab Duocarmazine)	20–40%, particularly high in serous carcinoma (up to ~80%)
HER3 (Patritumab Deruxtecan)	46.9 - 100.0 %

Fig. 10.4 Targeted tumor antigens of endometrial cancer for antibody-drug conjugates

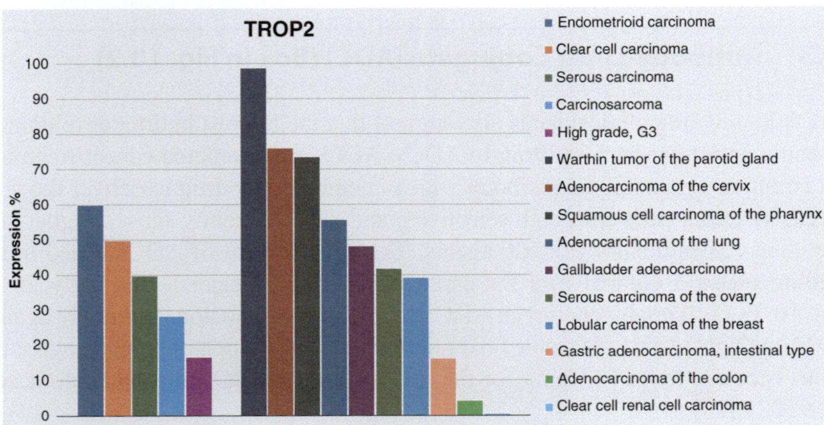

Fig. 10.5 Representative data on TROP2 expression in endometrial cancer

was approximately 60.1% in endometrial cancer and higher than in various cancers at other sites. That study also indicated that TROP2 expression was dependent on the histological subtype and was higher in low-grade endometrioid carcinoma than in carcinosarcoma and high-grade endometrioid carcinoma of special subtypes [32] (Fig. 10.5). FRα is a glycosylphosphatidylinositol-linked cell-surface glycoprotein that is widely expressed in certain cancers, but its expression is limited in normal tissues. The expression of FRα is found in 80–96%, 50.5–89.0%, and 67–80% of ovarian cancer, endometrial cancer, and triple-negative breast cancer, respectively, and is higher than that in gastrointestinal cancer (33–44%) or head and neck cancer (up to 45%) [33–36]. HER3 is a member of the EGFR family of RTKs. When HER3 is activated in a ligand-dependent or ligand-independent manner, it can bind to other receptors (the most potent binding partner is HER2) to regulate cell functions, including their growth and survival, nutrient sensing, and metabolic regulation through the PI3K–AKT–mTOR pathway [37]. HER3 has been found to promote tumorigenesis, tumor growth, and drug resistance in different cancer types and is frequently overexpressed in endometrial cancer (up to 100%) [38, 39]. The representative conjugated payloads of ADCs are DNA topoisomerase I inhibitors, including SN38 of an active irinotecan metabolite, belotecan derivative, and deruxtecan, and microtubule-targeting agents, including eribulin and SC209.

Sacituzumab tirumotecan (SKB264/MK-2870) is a TROP2 ADC developed using a novel linker to conjugate a belotecan-derivative topoisomerase I inhibitor. Sacituzumab tirumotecan demonstrated a 69% reduction in the risk of progression or death compared to conventional cytotoxic agents, and the objective response rate was 43.8% in a phase III clinical trial for patients with previously treated locally recurrent or metastatic triple-negative breast cancer. The most common grade ≥3 treatment-related adverse events were a decreased neutrophil count (32.3%), anemia (27.7%), and a decreased white blood cell count (25.4%) [40]. Sacituzumab govitecan is an antibody-drug conjugate composed of an antibody targeting Trop-2 coupled to SN-38 via a proprietary hydrolyzable linker. In a randomized phase III trial in patients with metastatic triple-negative breast cancer, the median OS was 12.1 months with sacituzumab govitecan and 6.7 months with a cytotoxic agent (HR for death, 0.48). In addition, the percentage of patients with an objective response was 35% to sacituzumab govitecan and 5% to chemotherapy. The incidence of key treatment-related adverse events of grade ≥3 were neutropenia (51%), leukopenia (10%), diarrhea (10%), anemia (8%), and febrile neutropenia (6%) [41]. The Food and Drug Administration (FDA) approved sacituzumab govitecan for patients with unresectable locally advanced or metastatic hormone receptor-positive, HER2-negative breast cancer who have received endocrine-based therapy and at least two additional systemic therapies in the metastatic setting.

In patients with endometrial cancer who experienced recurrence after platinum-based chemotherapy, the complete and partial response rates of ADCs ranged from 22.0% to 34.1% in phase II clinical trials [29, 42, 43] (NCT04205630), except for trastuzumab deruxtecan which demonstrated a confirmed objective response rate of 57.5% in patients with HER2-expressing endometrial cancer [44]. Therefore, when ADCs are used as a single-agent therapy, these response rates may be lower than single therapy in cancers of other sites or existing combination therapy with RTK inhibitors and ICIs for endometrial cancer. ADCs for endometrial cancer can be utilized using two different strategies. One is combination therapy with cytotoxic agents to achieve a maximal response. For example, brentuximab vedotin, which is a CD30-directed ADC linked to the cytotoxic payload monomethyl auristatin E, is used to treat Hodgkin lymphoma in combination with doxorubicin, vinblastine, and dacarbazine [45], and polatuzumab vedotin, which is an ADC targeting CD79b, is used to treat diffuse large B cell lymphoma in combination with rituximab, cyclophosphamide, and doxorubicin [46]. This strategy may be critical for patients with contraindications for ICIs, and one ongoing phase II study for ovarian cancer is investigating the effect of datopotamab deruxtecan with anticancer agents (NCT05489211). Another strategy is salvage treatment with a single-agent ADC in ICI-resistant patients. An ongoing phase III clinical trial is examining the impact of sacituzumab govitecan in treating patients with recurrent or persistent endometrial cancer, and eligible patients had previously undergone treatment with anti-PD-1 or PD-L1 antibodies (ASCENT-GYN-01/GOG-3104/ENGOT-en26, NCT06486441).

10.4 Alkylating Deacetylase Inhibitor

Tinostamustine is an alkylating histone deacetylase inhibitor (HDACi) fusion molecule composed of the alkylating agent bendamustine and pan-HDACi vorinostat, which has potential antineoplastic activity. Upon administration of tinostamustine, the vorinostat targets and binds to HDACs. This leads to the accumulation of highly acetylated histones, which results in the induction of chromatin remodeling, modulation of gene expression, inhibition of tumor cell division, and induction of tumor cell apoptosis. In a phase II study of tinostamustine, three of six of patients (50.0%) with advanced endometrial cancer demonstrated stable disease for more than 4 months [47]. Serious tinostamustine-related adverse events were reported in 30.6% of the patients; 27.8% discontinued treatment due to TEAEs, of which a decrease in the platelet count was the most common. Notably, one patient experienced a fatal TEAE (intra-abdominal hemorrhaging) [47].

10.5 Hormonal Therapy

A phase II randomized controlled clinical trial that enrolled patients with histologically confirmed clinically stage 1 FIGO grade 1 endometrial cancer demonstrated that intrauterine levonorgestrel, with or without metformin or weight loss, achieved an overall pathological complete response of 43% after 6 months [48]. In this study, the incidence of grade 3 adverse effects at 6 months was 4.2%, with infectious disease being the most frequent. In Japan and several other countries, the levonorgestrel-releasing intrauterine system (LNG-IUS) is labeled as contraindicated in patients with uterine malignancy, including endometrial carcinoma. Accordingly, its use for atypical endometrial hyperplasia (AEH) or endometrial cancer constitutes off-label/investigational therapy and should be limited to clinical trials or managed under institutional protocols with informed consent.

10.6 Other Negative Studies

To establish novel treatments, we need to know the results of many trials that did not meet the primary endpoint but could still provide valuable information to identify what does not work, refine future research directions, and prevent wasted resources. This section introduces several recent negative clinical studies of drugs other than ICIs whose results should be kept in mind.

The PREMIUM study, which is a multicenter, randomized, double-blind, placebo-controlled phase III study, demonstrated that there was no overall difference in the post-treatment Ki-67 between the metformin and placebo arms, and short-term treatment with standard diabetic doses of metformin did not reduce tumor proliferation in women with endometrioid endometrial cancer scheduled to undergo surgical treatment by hysterectomy in the following 5–35 days [49].

A combination maintenance therapy of the anti-angiogenesis agent bevacizumab with the PARP inhibitor olaparib demonstrated a significant effect in advanced ovarian cancer with genome instability and has become a standard treatment [50]. An open-label, randomized, phase 2 trial (NRG-GY012, NCT03660826) investigated the efficacy of olaparib in combination with the antiangiogenic agent cediranib, which is an inhibitor of all three VEGFR tyrosine kinases (VEGFR-1, VEGFR-2, VEGFR-3), in patients with advanced endometrial cancer. The combination of cediranib and olaparib has demonstrated modest clinical efficacy. The median PFS for cediranib was 3.8 months compared with 2.0 months for olaparib and 5.5 months for olaparib with cediranib. However, investigators reported that the primary endpoint was not met [51].

Zoptarelin (alternative names: AEZS-108, AN-152, ZEN-008) is a peptide agonist of the gonadotropin-releasing hormone-1 receptor (GnRH-1R) conjugated to the anthracycline antibiotic doxorubicin with potential antineoplastic activity. Zoptarelin doxorubicin binds to GnRH-1R, which is highly expressed on the endometrial and ovarian tumor cell membrane surfaces and is internalized. Once inside the cell, the doxorubicin moiety of this agent intercalates into the DNA and inhibits topoisomerase II activity, which may inhibit tumor cell DNA replication and proliferation. In an international multicenter phase III trial that registered patients with advanced, recurrent, or metastatic endometrial cancers who had failed prior platinum and taxane therapy, the median OS of patients treated with zoptarelin was 10.9 months compared to 10.8 months for patients treated with doxorubicin (HR = 1.06; 95% CI. 0.87, 1.30), and zoptarelin did not show a significant impact [52].

MK-2206 is an Akt inhibitor that binds to and inhibits the activity of the PI3K/Akt signaling pathway and inhibits tumor cell proliferation to induce tumor cell apoptosis. In a phase II trial in recurrent endometrial cancer, the objective response rate was only 6% [53].

The Lunchbox trial was a randomized phase III trial of cisplatin and irradiation, followed by carboplatin and paclitaxel vs. sandwich therapy of carboplatin and paclitaxel, followed by irradiation and then carboplatin and paclitaxel in patients with FIGO 2008 stage III or IVA endometrial carcinoma or stage I or II serous or clear cell endometrial carcinoma. Although this clinical trial was closed early owing to low accrual, the 3-year recurrence-free survival (RFS) was 85.7% in the control arm and 73.4% in the sandwich therapy group ($p = 0.58$), and sandwich therapy did not show a significant effect [54].

Arsenic trioxide is a small-molecule arsenic compound that damages or degrades the promyelocytic leukemia/retinoic acid receptor-alpha (PML/RARa) fusion protein. Arsenic trioxide induces apoptosis in acute promyelocytic leukemia cells and many other cancer cell types. In addition, arsenic trioxide promotes cell differentiation and exerts a pro-angiogenic effect. A phase II, open-label, single-arm clinical trial in women who had already received chemotherapy for endometrial cancer or who had a disease that had spread outside of the uterus showed no objective response (NCT01184053).

Nintedanib is an orally bioavailable, indolinone-derived inhibitor of multiple RTKs and non-RTKs. Upon administration, nintedanib selectively binds to and inhibits VEGFR, FGFR, PDGFR, and colony-stimulating factor 1 receptor (CSF1R) tyrosine kinases. A randomized phase II study of combination chemotherapy with nintedanib in advanced and recurrent endometrial cancer demonstrated that nintedanib added to chemotherapy did not improve the PFS compared to chemotherapy plus placebo: median for nintedanib vs. placebo of 8.3 vs. 7.2 months; HR adjusted for stratification factors 1.03; 95% CI 0.71–1.48 [55].

These results suggest that novel treatments targeting a single molecule are still challenging, and drugs with general purpose, such as anti-angiogenesis or immune checkpoint inhibition, are advantageous in patients with endometrial cancer.

10.7 Future Directions

The results of recent clinical trials suggest that developing novel treatments limited to endometrial cancer is quite difficult, although several of the treatments described above appear to be promising. In the future, clinical trials based on the genetic background of such patients are required across multiple primary sites that include endometrial cancer.

ICIs are the mainstay of novel treatments for advanced and recurrent endometrial cancer. However, approximately half of recurrent or refractory patients do not respond to ICIs, and their combination therapy and serious immune-related adverse events prevent some patients from continuing further treatments, even if they are effective. In addition, cancer genome profiling (CGP) tests for targeted therapy are still being developed, and the rate of individualized treatments with strong recommendations is approximately 10% in Japan [56] because there are fewer patients with key genetic alterations that include *TP53* and *HER2* than expected and fewer pragmatic drugs targeting representative genetic alterations. Therefore, novel salvage treatments are still required.

The results of previous clinical trials introduced in this chapter suggest that effective drugs for endometrial cancer are likely to have a global effect on the tumor microenvironment, such as ICIs, ADCs equipped with a payload that has a broad cytotoxic effect, and angiogenesis inhibitors, rather than targeted drugs against single molecules in the signaling pathway associated with cancer cell proliferation. Hormonal therapy should have a broad potential to inhibit the growth of hormone receptor-positive endometrial cancer, but the benefits are still limited. CGP tests to search for single-molecule targeted agents currently do not seem sufficiently useful in searching for therapeutic drugs with a high response in patients with endometrial cancer. A paradigm shift may occur, moving from the current CGP to searching for limited actionable alterations in hundreds of genes via a comprehensive whole-genome analysis. We may then be able to expect higher therapeutic actionability and the discovery of novel target molecules useful for treating endometrial cancer [57]. In addition, recent evidence suggests that artificial intelligence (AI) can help to create new medications and treatments, improve early cancer detection through image analyses, and generate ultimate individualized treatment plans based on patient information (Fig. 10.2), as AI takes advantage of large language models to develop new computational tools to improve knowledge extraction from research publications. In the future, we need to verify the introduction and application of cancer genome profiling using whole-genome sequencing and AI systems for treatment individualization in each region.

ADCs are expected to become primary candidates for new treatments for endometrial cancer. In addition, it is also noteworthy that progress has been made both in making efficient use of existing antiangiogenic agents and in developing new drugs targeting p53, which has been a long-standing goal since previous studies have indicated that p53 mutations occur in endometrial cancer. We desire the treatment development described above to improve the prognosis of patients, especially in heavily pretreated patient populations with advanced and metastatic endometrial cancer.

References

1. Thigpen JT, Brady MF, Homesley HD, et al. Phase III trial of doxorubicin with or without cisplatin in advanced endometrial carcinoma: a gynecologic oncology group study. J Clin Oncol. 2004;22(19):3902–8.
2. Miller DS, Filiaci VL, Mannel RS, et al. Carboplatin and paclitaxel for advanced endometrial cancer: final overall survival and adverse event analysis of a phase III trial (NRG oncology/GOG0209). J Clin Oncol. 2020;38(33):3841–50.
3. Cancer Genome Atlas Research Network, Kandoth C, Schultz N, et al. Integrated genomic characterization of endometrial carcinoma. Nature. 2013;497(7447):67–73.
4. Pellot Ortiz KI, Rechberger JS, Nonnenbroich LF, Daniels DJ, Sarkaria JN. MDM2 inhibition in the treatment of glioblastoma: from concept to clinical investigation. Biomedicine. 2023;11(7):1879.
5. Jeczen R, Skomra D, Cybulski M, et al. P53/MDM2 overexpression in metastatic endometrial cancer: correlation with clinicopathological features and patient outcome. Clin Exp Metastasis. 2007;24(7):503–11.
6. MKK W, Burgess MA, Chandra S, et al. Navtemadlin (KRT-232) activity after failure of anti-PD-1/L1 therapy in patients (pts) with *TP53WT* Merkel cell carcinoma (MCC). J Clin Oncol. 2022;40(16_suppl):9506.
7. National Center for Biotechnology Information. PubChem compound summary for CID 25017411, anlotinib. 2025. https://pubchem.ncbi.nlm.nih.gov/compound/Anlotinib.
8. Shen G, Zheng F, Ren D, et al. Anlotinib: a novel multi-targeting tyrosine kinase inhibitor in clinical development. J Hematol Oncol. 2018;11(1):120.
9. Li S, Wang H. Research progress on mechanism and management of adverse drug reactions of anlotinib. Drug Des Devel Ther. 2023;17:3429–37.
10. Han B, Li K, Wang Q, et al. Effect of anlotinib as a third-line or further treatment on overall survival of patients with advanced non-small cell lung cancer: the ALTER 0303 phase 3 randomized clinical trial. JAMA Oncol. 2018;4(11):1569–75.
11. Syed YY. Anlotinib: first global approval. Drugs. 2018;78(10):1057–62.
12. Cui Q, Mao Y, Hu Y, Ma D, Liu H. Anlotinib in recurrent or metastatic endometrial cancer. Int J Gynecol Cancer. 2022;32(9):1147–52.
13. Aghajanian C, Sill MW, Darcy KM, et al. Phase II trial of bevacizumab in recurrent or persistent endometrial cancer: a gynecologic oncology group study. J Clin Oncol. 2011;29(16):2259–65.
14. Hack SP, Zhu AX, Wang Y. Augmenting anticancer immunity through combined targeting of angiogenic and PD-1/PD-L1 pathways: challenges and opportunities. Front Immunol. 2020;11:598877.

15. Makker V, Colombo N, Casado Herraez A, et al. Lenvatinib plus pembrolizumab for advanced endometrial cancer. N Engl J Med. 2022;386(5):437–48.
16. Vergote I, Perez-Fidalgo JA, Hamilton EP, et al. Oral selinexor as maintenance therapy after first-line chemotherapy for advanced or recurrent endometrial cancer. J Clin Oncol. 2023;41(35):5400–10.
17. Makker V, Perez-Fidalgo JA, Valabrega G, et al. Long-term follow-up of efficacy and safety of selinexor maintenance treatment in patients with TP53wt advanced or recurrent endometrial cancer: a subgroup analysis of the ENGOT-EN5/GOG-3055/SIENDO study. Gynecol Oncol. 2024;185:202–11.
18. Leary A, Estevez-Garcia P, Sabatier R, et al. ENDOLUNG trial. A phase 1/2 study of the Akt/mTOR inhibitor and autophagy inducer Ibrilatazar (ABTL0812) in combination with paclitaxel/carboplatin in patients with advanced/recurrent endometrial cancer. BMC Cancer. 2024;24(1):876.
19. Slomovitz BM, Filiaci VL, Walker JL, et al. A randomized phase II trial of everolimus and letrozole or hormonal therapy in women with advanced, persistent or recurrent endometrial carcinoma: a GOG foundation study. Gynecol Oncol. 2022;164(3):481–91.
20. Heudel P, Frenel JS, Dalban C, et al. Safety and efficacy of the mTOR inhibitor, vistusertib, combined with anastrozole in patients with hormone receptor-positive recurrent or metastatic endometrial cancer: the VICTORIA Multicenter, open-label, phase 1/2 randomized clinical trial. JAMA Oncol. 2022;8(7):1001–9.
21. Han SN, Oza A, Colombo N, et al. A randomized phase 2 study of sapanisertib in combination with paclitaxel versus paclitaxel alone in women with advanced, recurrent, or persistent endometrial cancer. Gynecol Oncol. 2023;178:110–8.
22. Fader AN, Roque DM, Siegel E, et al. Randomized phase II trial of carboplatin-paclitaxel versus carboplatin-paclitaxel-trastuzumab in uterine serous carcinomas that overexpress human epidermal growth factor receptor 2/neu. J Clin Oncol. 2018;36(20):2044–51.
23. Fader AN, Roque DM, Siegel E, et al. Randomized phase II trial of carboplatin-paclitaxel compared with carboplatin-paclitaxel-trastuzumab in advanced (stage III-IV) or recurrent uterine serous carcinomas that overexpress Her2/Neu (NCT01367002): updated overall survival analysis. Clin Cancer Res. 2020;26(15):3928–35.
24. Janku F, Hamilton EP, Mathews CA, et al. Open-label, multicenter, phase 1b/2 study of rebastinib in combination with paclitaxel to assess safety and efficacy in patients with advanced or metastatic endometrial cancer. J Clin Oncol. 2021;39(15_suppl):5576.
25. Lorusso D, Ferrandina G, Colombo N, et al. Carboplatin-paclitaxel compared to carboplatin-paclitaxel-bevacizumab in advanced or recurrent endometrial cancer: MITO END-2 – a randomized phase II trial. Gynecol Oncol. 2019;155(3):406–12.
26. Leslie KK, Filiaci VL, Mallen AR, et al. Mutated p53 portends improvement in outcomes when bevacizumab is combined with chemotherapy in advanced/recurrent endometrial cancer: An NRG oncology study. Gynecol Oncol. 2021;161(1):113–21.
27. Thiel KW, Devor EJ, Filiaci VL, et al. TP53 sequencing and p53 immunohistochemistry predict outcomes when bevacizumab is added to frontline chemotherapy in endometrial cancer: An NRG Oncology/Gynecologic Oncology Group Study. J Clin Oncol. 2022;40(28):3289–300.
28. Deeks ED. Disitamab vedotin: first approval. Drugs. 2021;81(16):1929–35.
29. Santin AD, Corr BR, Spira A, et al. Efficacy and safety of sacituzumab govitecan in patients with advanced solid tumors (TROPiCS-03): analysis in patients with advanced endometrial cancer. J Clin Oncol. 2024;42(29):3421–9.
30. Bignotti E, Zanotti L, Calza S, et al. Trop-2 protein overexpression is an independent marker for predicting disease recurrence in endometrioid endometrial carcinoma. BMC Clin Pathol. 2012;12:22.
31. Celik SY, Celik OI. Can TROP2 be used as a prognostic marker in endometrioid endometrial carcinoma? Indian J Pathol Microbiol. 2020;63(3):418–22.
32. Dum D, Taherpour N, Menz A, et al. Trophoblast cell surface antigen 2 expression in human tumors: a tissue microarray study on 18,563 Tumors. Pathobiology. 2022;89(4):245–58.

33. Gonzalez T, Muminovic M, Nano O, Vulfovich M. Folate receptor alpha-a novel approach to cancer therapy. Int J Mol Sci. 2024;25(2):1046.
34. Senol S, Ceyran AB, Aydin A, et al. Folate receptor alpha expression and significance in endometrioid endometrium carcinoma and endometrial hyperplasia. Int J Clin Exp Pathol. 2015;8(5):5633–41.
35. O'Shannessy DJ, Somers EB, Smale R, Fu YS. Expression of folate receptor-alpha (FRA) in gynecologic malignancies and its relationship to the tumor type. Int J Gynecol Pathol. 2013;32(3):258–68.
36. Dainty LA, Risinger JI, Morrison C, et al. Overexpression of folate binding protein and mesothelin are associated with uterine serous carcinoma. Gynecol Oncol. 2007;105(3):563–70.
37. Majumder A. HER3: toward the prognostic significance, therapeutic potential, current challenges, and future therapeutics in different types of cancer. Cells. 2023;12(21):2517.
38. Kojima Y, Sudo K, Yoshida H, et al. Changes in HER3 expression profiles between primary and recurrent gynecological cancers. Cancer Cell Int. 2023;23(1):18.
39. Androutsopoulos G, Adonakis G, Liava A, Ravazoula P, Decavalas G. Expression and potential role of ErbB receptors in type II endometrial cancer. Eur J Obstet Gynecol Reprod Biol. 2013;168(2):204–8.
40. Xu B, Yin Y, Fan Y, et al. Sacituzumab tirumotecan (SKB264/MK-2870) in patients (pts) with previously treated locally recurrent or metastatic triple-negative breast cancer (TNBC): results from the phase III OptiTROP-Breast01 study. J Clin Oncol. 2024;42(16_suppl):104.
41. Bardia A, Hurvitz SA, Tolaney SM, et al. Sacituzumab govitecan in metastatic triple-negative breast cancer. N Engl J Med. 2021;384(16):1529–41.
42. Wang D, Wang K, An R, et al. Safety and efficacy of sacituzumab tirumotecan (sac-TMT) in patients (pts) with previously treated advanced endometrial carcinoma (EC) and ovarian cancer (OC) from a phase II study. Ann Oncol. 2024;35(suppl_2):S544–S95.
43. Oaknin A, Ang JE, Rha SY, et al. 714MO Datopotamab deruxtecan (Dato-DXd) in patients with endometrial (EC) or ovarian cancer (OC): results from the phase II TROPION-PanTumor03 study. Ann Oncol. 2024;35:S547–S8.
44. Meric-Bernstam F, Makker V, Oaknin A, et al. Efficacy and safety of trastuzumab deruxtecan in patients with HER2-expressing solid Tumors: primary results from the DESTINY-PanTumor02 phase II trial. J Clin Oncol. 2024;42(1):47–58.
45. Ansell SM, Radford J, Connors JM, et al. Overall survival with brentuximab vedotin in stage III or IV Hodgkin's lymphoma. N Engl J Med. 2022;387(4):310–20.
46. Tilly H, Morschhauser F, Sehn LH, et al. Polatuzumab vedotin in previously untreated diffuse large B-cell lymphoma. N Engl J Med. 2022;386(4):351–63.
47. Tinker A, Strauss JF, Schneider R, et al. A phase II study of tinostamustine in patients (pts) with advanced solid tumours. J Clin Oncol. 2023;41(16_suppl):3021.
48. Janda M, Robledo KP, Gebski V, et al. Complete pathological response following levonorgestrel intrauterine device in clinically stage 1 endometrial adenocarcinoma: results of a randomized clinical trial. Gynecol Oncol. 2021;161(1):143–51.
49. Kitson SJ, Maskell Z, Sivalingam VN, et al. PRE-surgical metformin in uterine malignancy (PREMIUM): a multi-center, randomized double-blind, placebo-controlled phase III trial. Clin Cancer Res. 2019;25(8):2424–32.
50. Ray-Coquard I, Leary A, Pignata S, et al. Olaparib plus bevacizumab first-line maintenance in ovarian cancer: final overall survival results from the PAOLA-1/ENGOT-ov25 trial. Ann Oncol. 2023;34(8):681–92.
51. Rimel BJ, Enserro D, Bender DP, et al. NRG-GY012: randomized phase 2 study comparing olaparib, cediranib, and the combination of cediranib/olaparib in women with recurrent, persistent, or metastatic endometrial cancer. Cancer. 2024;130(8):1234–45.
52. Miller DS, Scambia G, Bondarenko I, et al. ZoptEC: phase III randomized controlled study comparing zoptarelin with doxorubicin as second line therapy for locally advanced, recurrent, or metastatic endometrial cancer (NCT01767155). J Clin Oncol. 2018;36(15_suppl):5503.

53. Myers AP, Konstantinopoulos PA, Barry WT, et al. Phase II, 2-stage, 2-arm, PIK3CA mutation stratified trial of MK-2206 in recurrent endometrial cancer. Int J Cancer. 2020;147(2):413–22.
54. Barlin JN, Mahar B, Ata A, et al. Lunchbox trial: a randomized phase III trial of cisplatin and irradiation followed by carboplatin and paclitaxel versus sandwich therapy of carboplatin and paclitaxel followed by irradiation then carboplatin and paclitaxel for advanced endometrial carcinoma. Gynecol Oncol. 2024;180:63–9.
55. Mirza M, Berton D, Vergote I, et al. A randomised phase II study of combination chemotherapy with nintedanib/placebo in advanced/recurrent endometrial cancer. FANDANGO/ENGOT-EN1/FANDANGO. Int J Gynecol Cancer. 2021;31:A371–A2.
56. Hagio K, Kikuchi J, Takada K, et al. Assessment for the timing of comprehensive genomic profiling tests in patients with advanced solid cancers. Cancer Sci. 2023;114(8):3385–95.
57. Kim R, Kim S, Oh BB, et al. Clinical application of whole-genome sequencing of solid tumors for precision oncology. Exp Mol Med. 2024;56(8):1856–68.

Fertility-Preserving Treatment for Endometrial Cancer

11

Shoko Kitazawa, Kensuke Sakai, and Wataru Yamagami

Abstract

Endometrial cancer is one of the most common gynecological malignancies and is increasingly diagnosed in younger women, highlighting the growing need for fertility-preserving treatments. Progestin therapy using agents, such as medroxyprogesterone acetate and megestrol acetate, offers a non-surgical option for managing atypical endometrial hyperplasia and early-stage endometrial cancer in patients who desire to conceive.

This chapter explores the efficacy and challenges of progestin therapy with complete response rates between 67% and 98.5%. Combination approaches, such as those incorporating levonorgestrel-releasing intrauterine devices or metformin, show promise for improving treatment outcomes. However, recurrence remains a significant concern, particularly in patients with endometrioid carcinoma grade 1. Genomic factors, including microsatellite instability-high/mismatch repair deficient status and Lynch syndrome, may further influence treatment success and risk of recurrence.

Although progestin therapy provides a critical option for fertility preservation, careful patient selection, close monitoring for complications, such as thrombosis and liver dysfunction, and prompt transition to infertility treatment are essential. This chapter emphasizes the need for continued studies on the molecular subtypes and genetic predictors to refine treatment strategies and improve long-term outcomes in patients undergoing fertility-sparing treatment.

S. Kitazawa · K. Sakai · W. Yamagami (✉)
Keio University School of Medicine, Department of Obstetrics and Gynecology, Tokyo, Japan
e-mail: k_sakai@keio.jp; gami.z8@keio.jp

H. Watari, T. Mitamura (eds.), *Uterine Body Neoplasms*, Comprehensive Gynecology and Obstetrics, https://doi.org/10.1007/978-981-95-0946-1_11

Keywords

Fertility-preserving treatment · Endometrial cancer · Progestin therapy

11.1 Introduction

Endometrial cancer is the sixth most frequently diagnosed malignancy in terms of incidence for the year 2022 [1]. Relative excess estrogen and persistence of unopposed estrogen contribute to the carcinogenesis of endometrial cancer, especially in high-income countries where Western dietary habits, such as high-fat diets, are prevalent. The prevalence of endometrial cancer is increasing as patient population is getting younger. Additionally, the increasing incidence of non-communicable diseases is an issue in low- and middle-income countries [2], and in the future, the incidence of endometrial cancer is expected to increase in developing countries.

Although the standard treatment for atypical endometrial hyperplasia (AEH) and early-stage endometrial cancer is surgery known as hysterectomy, it is critical to preserve the uterus in those who strongly desire a baby. Considering the effects of excess estrogen, progestin therapy is considered for endometrial cancer. Medroxyprogesterone acetate (MPA) and megestrol acetate (MA) are commonly used as oral medicines, either alone or in combination with metformin and have been reported to be beneficial in some cases. Moreover, levonorgestrel-releasing intrauterine devices (LNG-IUDs) and gonadotropin-releasing hormone analogs (GnRHa) are used in some patients [3, 4]. Although these treatments are expected to be effective against endometrial cancer, some are known to have a risk of side effects such as thrombosis [5]. Although progestin therapy is indicated for the treatment of early-stage endometrial cancer, we must keep in mind that it is a treatment for malignant tumors and must be carefully evaluated and managed. Furthermore, there have been only a few reports on genome-related factors that may be one of the carcinogenic mechanisms of endometrial cancer in recent years, suggesting that fertility-preserving treatments for endometrial cancer are still unclear. Here, we discuss fertility preservation therapy for AEH and early-stage endometrial cancer, especially endometrioid carcinoma grade 1 (EMG1) for those who desire to have a baby.

11.2 Overview: Progestin Therapy as Fertility-Preserving Treatment for AEH and Early-Stage Endometrial Cancer

11.2.1 Progestin as a Fertility-Sparing Treatment (P-FST)

High-dose progestins are used to counter the state of unopposed estrogen. The European Society of Gynecological Oncology/European Society for Radiotherapy and Oncology/European Society of Pathology and National Comprehensive Cancer

Network (NCCN) guidelines recommend MPA (400–600 mg/day) or MA (160–320 mg/day) as an oral medicine [6–8]. In addition to them, LNG-IUDs or levonorgestrel-releasing intrauterine systems (LNG-IUSs), and GnRHa have also been used. MPA is covered by health insurance for AEH and EMG1 in Japan, making it the most widely used treatment. However, other agents, such as MA, LNG-IUDs, and GnRHa, may also be used in other countries [6, 7]. One report indicated that combination therapy with GnRHa and LNG-IUDs achieved a high complete response (CR) rate within 26 weeks [3].

Other progestins, including hydroxyprogesterone caproate, letrozole, norethisterone acetate, natural progesterone, and dydrogesterone, are also used, but are not widely recommended for fertility-preserving treatment [9, 10].

11.2.2 Non-surgical Treatment Options

The most widely used P-FST treatments are MPA and MA. Both MPA and MA demonstrate high response rates in P-FST and are considered reliable therapeutic options. Although one study highlighted that MA has been shown to result in higher remission rates than MPA and other hormonal treatments [11], possibly due to its relatively higher bioavailability following oral administration, there is no strong evidence or reports suggesting significant differences in CR rates or pregnancy rates between the two medications [7]. Generally, it is widely considered that MPA is more preferable than MA. In terms of side effects, one study reported that MA was associated with a significantly higher rate of weight gain, affecting 43.1% of patients [12].

Additionally, LNG-IUSs have shown promise as an alternative to oral medications. The LNG-IUS is a T-shaped device inserted into the uterus and is commonly used as a contraceptive or treatment for abnormal bleeding [13]. They can also be combined with oral contraceptives or GnRHa. Some studies have reported that combining LNG-IUSs with GnRHa is effective [3, 4], whereas others suggest that LNG-IUS alone is non-inferior to a combination of LNG-IUS and MA [14]. There are also reports that the combination of LNG-IUS and MPA has shown high effectiveness [15–17]. Moreover, LNG-IUSs have fewer side effects, such as weight gain, nausea, and liver dysfunction, which are commonly associated with oral medications. Once inserted, the effect of progesterone persists unless the device is removed, eliminating concerns regarding patient compliance.

The choice of treatment may depend on the patient's side effects and ability to adhere to oral medication. However, caution should be exercised when using LNG-IUSs for P-FST. Although endometrial biopsies are doable, the accuracy of the examination is not guaranteed, and total curettage and hysteroscopy may be challenging. These factors require careful consideration when selecting LNG-IUSs for P-FST [14, 18].

11.2.3 Metformin

In addition to the progestin mentioned above, it is expected that the combination of metformin may increase the efficacy of P-FST. Metformin is an oral biguanide hypoglycemic medication primarily used in patients with type 2 diabetes. Recently, it was shown to have anti-tumor effects in hormone-dependent tumors, especially uterine and breast cancers [19]. This is thought to be mediated by metformin's inhibition of the mTOR signaling pathway, promotion of apoptosis, and steroid receptor modulation [20, 21]. Many patients seeking P-FST are obese and insulin resistant, which are risk factors for diabetes and metabolic syndrome [22]. In our institution, out of 319 cases, 50 had a BMI of 30 or higher, and 77 had a BMI of 22 or higher. Mitsuhashi et al. reported that 82% of patients undergoing P-FST with metformin achieved CR within 36 weeks, and only 10.3% relapsed after achieving CR, which is a much lower relapse rate [20]. This result does not necessarily mean that P-FST with metformin lowers the remission rate, but reduces the possibility of recurrence. Furthermore, metformin inhibits weight gain, which may be a side effect of P-FST (see Sect. 11.4) [21]. However, no severe toxicity was observed. Therefore, the FELICIA trial is currently underway to investigate the effects of combining metformin with P-FST in Japan [22]. Polycystic ovary is also considered to be a cause of endometrial cancer. Metformin is also expected to promote ovulation and improve pregnancy rates in patients with polycystic ovarian syndrome, suggesting a synergistic effect [20, 22].

11.2.4 Aspirin

As mentioned above, individuals at a high risk of endometrial cancer are often obese. Obesity-related atherothrombotic risk is primarily attributed to a prothrombotic state characterized by elevated levels of coagulation factors, impaired fibrinolysis, and heightened platelet reactivity. These conditions result from an interplay between factors commonly associated with obesity, including insulin resistance, inflammation, oxidative stress, and endothelial dysfunction.

Aspirin, an antiplatelet agent, was expected to reduce the incidence and mortality of certain cancers, particularly colorectal cancer [23, 24]. This effect is believed to result from the ability of aspirin to inhibit metastasis by inactivating platelets and suppressing prostaglandin E2 [25]. There are also reports suggesting that aspirin may reduce the risk of endometrial cancer, making its use in combination with P-FST a potential option [26]. However, the evidence supporting this approach remains limited. Regarding colorectal cancer, the results of the ASPREE trial pointed out that considering the side effects of aspirin, the use of low-dose aspirin in elderly individuals (aged 70 and older) may actually increase cancer mortality. In response to these findings, the U.S. Preventive Services Task Force (USPSTF) no longer recommends aspirin use for colorectal cancer prevention [27].

Moreover, the use of aspirin raises concerns, such as an increased risk of irregular genital bleeding, which is a common symptom of endometrial cancer. Other considerations include the timing of aspirin discontinuation prior to procedures such as dilation and curettage (D&C). Although there is optimism regarding its potential benefits, aspirin is not currently regarded as a suitable medication for routine prescription in this context.

11.2.5 Criteria and Management for P-FST

Considering that the standard treatment for AEH and endometrial cancer is hysterectomy and that progestin is a temporary treatment to preserve fertility, the indication for P-FST using progestin must be carefully determined. No standardized criteria or management methods have been established for P-FST. However, progestin therapy is generally administered, and because intensive follow-up is required to assess the endometrial response, most authors recommend endometrial sampling every 3–6 months, either through D&C or hysteroscopic biopsy [7]. Fertility-sparing treatment should be continued until histological confirmation of lesion disappearance, which is called complete remission. Although D&C under anesthesia is preferred for histological evaluation, the usefulness of hysteroscopy has been increasingly highlighted and will be discussed in detail in another paragraph. No randomized controlled trials have established a clear and strict interval or assessment method for the follow-up of patients with endometrial cancer after fertility preservation.

The following are the indications for P-FST at our hospital [37], which performs one of the most common P-FST procedures in Japan. This is just our hospital's method and is not a generally recommended protocol, but we will introduce the criteria for the MPA therapy we use. High-dose MPA therapy is recommended for young patients with a strong desire to conceive (generally age 42 years or younger at our hospital) who have AEH or EMG1, no myometrial invasion on pelvic MRI, and no evidence of extrauterine lesions on thoracoabdominal pelvic computed tomography (CT). Both imaging modalities should be evaluated whenever possible. Although AEH is a precancerous condition, it may be diagnosed as malignant cancer during tests such as endometrial biopsy or D&C, or after hysterectomy. To rule out those possibilities, CT is useful for its diagnosis because of the potential risk of malignant complications.

In addition to these major criteria, our hospital considers the following minor criteria to determine eligibility: age (≤42 years), body mass index (BMI) of <35 kg/m^2, normal liver function, no coagulation abnormalities, no history of thrombosis, nonsmoking status, and ability to attend regular hospital visits (Fig. 11.1).

If MPA therapy is deemed appropriate after histological and imaging diagnoses via D&C, treatment is initiated with MPA (HyslonH tablet®) at 600 mg/day. Following the start of therapy, patients underwent monthly evaluations, including a medical interview, internal examination, transvaginal ultrasound, endometrial cytology, endometrial biopsy, and blood tests. Medication compliance and adverse events were assessed during medical interviews. Internal examination confirms uterine size and absence of bilateral adnexal enlargement. Transvaginal ultrasound was

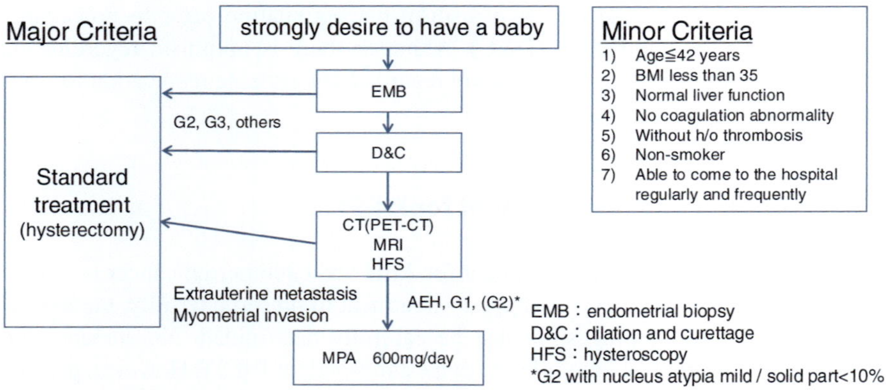

Fig. 11.1 Criteria for progestin therapy as fertility-preserving treatment for AEH and early-stage endometrial cancer at our institution

used to monitor endometrial thickening and check for abnormalities in the adnexa or the presence of ascites. Blood tests included liver enzyme levels, coagulation status, and tumor markers such as CA125.

Hysteroscopy is always performed before D&C to assess residual lesions and their localization. After at least 4 months of MPA treatment, D&C was performed to evaluate the endometrium. If D&C does not identify lesions more advanced than endometrial hyperplasia, the treatment was discontinued. If significant lesions persisted, therapy was continued and D&C was repeated every 2 months. Cytological diagnosis may be attempted during D&C to enhance diagnostic accuracy; however, its effectiveness is limited because pathological diagnosis remains the gold standard [28]. If persistent endometrial thickening or suspected extrauterine lesions are identified on transvaginal ultrasound, further imaging is required to confirm the absence of myometrial invasion or extrauterine metastasis (Fig. 11.2).

Once D&C confirms lesion resolution, patients desiring pregnancy were promptly referred to our hospital's outpatient reproductive infertility clinic for fertility treatment. Outpatients should be monitored every 3–4 months within the first 2 years and every 6 months thereafter. Recurrence can occur even after 10 years; therefore, follow-up continues beyond a decade unless the patient transitions towards standard treatment involving hysterectomy. If recurrence is suspected during follow-up and confirmed by D&C, and the patient still strongly wishes to preserve fertility and meets the above criteria, MPA therapy will be re-initiated.

This section outlines the general approach and protocol for P-FST implemented at our institution. However, there are other reported methods for performing P-FST, which are discussed in a separate section.

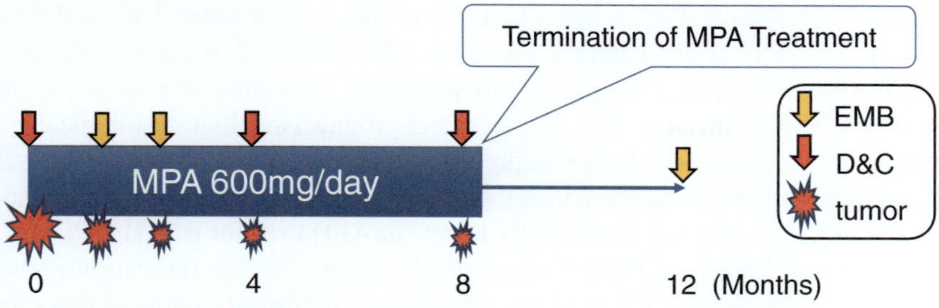

Fig. 11.2 Treatment protocol at Keio University Hospital

11.2.6 Hysteroscopy

The usefulness of hysteroscopy combined with D&C has been widely reported. First, hysteroscopy provides direct visualization, allowing targeted biopsies and resection of suspicious tissues, offering better diagnostic accuracy and lower risks than D&C alone [29–32]. Additionally, hysteroscopic tumor resection combined with hormonal therapy has been shown to achieve higher response rates and better pregnancy outcomes than hormonal therapy alone [33]. Hysteroscopic resection can enhance the effectiveness of P-FST by not only enabling detailed observation of the uterine cavity but also removing partial lesions that are otherwise difficult to eliminate.

Concerns have been raised about the possibility that large amounts of water introduced into the uterus during the procedure may allow cancer cells to flow into the abdominal cavity. Dong et al. reported that more than 80 mmHg of intrauterine pressure may contribute to intraperitoneal dissemination of malignant cells, but not with stages I–II [34]. Some reports have mentioned that this contributes to the possibility of cancer cell flow into the abdomen, which may result in positive peritoneal cytology [34]. However, there was no significant difference in positive peritoneal cytology in patients with endometrial cancer stages stages I [34]. Aumiphin et al. showed that hysteroscopy does not influence the stages of endometrial cancer or cause retrograde seeding in endometrial cancer [35].

11.3 Treatment Outcomes and Recurrence Rates

Fertility-sparing treatment for AEH and early endometrial cancer results in a high CR rate of 67–98.5% [36–44], and is thus considered to be highly effective. Additionally, the CR rate is significantly higher for AEH than for EMG1 [37, 41].

On the other hand, P-FST has a high recurrence rate between 16% and 63.2% [36–44]. As P-FST is not a standard treatment, it is essential to consider the possibility of recurrence under appropriate management. The endometrium is the most common site of recurrence. Factors that affect prognosis include histological type, whether or not the patient has been pregnant after P-FST, duration taken to reach CR, and whether or not maintenance therapy is given after treatment [37]. In addition, the recurrence rate is significantly lower for AEH than for EMG1 [37, 41].

Furthermore, repeated P-FST is acceptable when a similar indication remains applicable. Although the CR rate was also high in recurrent treatment outcomes, the recurrence rate of retreatment was higher than that of the initial treatment [44–49]. Yamagami et al. showed that the recurrence rate among patients who underwent repeated treatment for EMG1 was 81.1%, which was significantly higher than that during the initial treatment (63.2%). Moreover, they also pointed out that recurrence-free survival (RFS) in retreatment was significantly worse than that in the initial treatment [41]. Interestingly, based on patients at our institution, in patients where EMG1 was diagnosed at the time of recurrence, even if AEH was diagnosed at the time of initial treatment, the prognosis thereafter was due to EMG1. As this is not a standard treatment, it is necessary to decide how far to continue with the treatment for recurrence based on the histological type at the time of recurrence, patient's desire to become pregnant, course of treatment, etc., while fully discussing the matter with the patient, and if necessary, the patient's family. The REMPA trial, which is a multicenter phase II study, involves performing P-FST, specifically with MPA therapy, in patients with recurrence [50]. The REMPA trial is currently being conducted to evaluate the efficacy and safety of the treatment in patients who experience relapse after fertility preservation therapy for AEH or early-stage endometrial cancer. Participants were recruited from multiple facilities in Japan and South Korea.

No significant differences were observed in the CR rates among different treatment options for relapse, including MPA, MA, and LNG-IUS. Additionally, one study reported that 90.2% of patients achieved CR with treatment options primarily utilizing GnRH [49]. Furthermore, evidence suggests that maintenance therapy may be effective for patients who do not desire to conceive immediately after achieving CR [41].

As mentioned in Sect. 11.2, the indication for P-FST is AEH or EMG1 in younger generations who wish to preserve their fertility; however, there have also been reports of cases where P-FST was performed for other indications. For example, P-FST was performed with careful management for grade 2 patients, resulting in 54 of 84 patients achieving CR, while 20 experienced recurrences [51]. Additionally, there are reports indicating that 22 patients became pregnant and 8 gave birth to live babies [48]. In the Gorilla study, which evaluated the efficacy and safety of P-FST for stage I grade 2 and grade 1–2 endometrial cancer with shallow myometrial invasion, the CR rate was 72.2%, progressive disease (PD) rate was 16.6%, and recurrence rate was 38.5% [52, 53]. The study also reported that the risk of PD was significantly higher in cases where the tumor size prior to treatment was $\geq$2 cm.

11.4 Complications

The most frequent adverse events associated with P-FST are weight gain and liver dysfunction. Although most of these events were minor, grade 3 or higher adverse events occurred in 4.4% and 2.2% of patients, respectively [36]. Other adverse events included irregular genital bleeding, impaired glucose tolerance, and edema. Serious adverse events including thrombosis, heart failure, and anaphylaxis occur, although their frequencies are unknown.

Of particular concern is the risk of thrombosis from hormonal therapy in patients with obesity, with a reported relative risk of 6.10 (95% confidence interval [CI]: 6.04–6.17) for deep venous thrombosis in women with obesity aged <40 years and a BMI of 25–29.9 kg/m^2, 22/10,000 for BMI 30–34.9 kg/m^2, and 36.8/10,000 for BMI >35 kg/m^2, indicating that obesity increases the risk of thromboembolism (venous thromboembolism: VTE) [5]. Progestin itself is also a hormone product with a risk of thrombosis, and the odds ratio for VTE at high doses (≥20 mg/day) has been reported to be 4.00 (99% CI: 1.42–11.3) [54, 55]. Patients at risk of endometrial cancer tend to be obese in the first place and are, therefore, naturally at risk of thrombosis even if P-FST is indicated. Therefore, considering the risk of thrombosis, patients with a BMI of <35 kg/m^2 are indicated in our hospital. Aspirin may prevent thrombotic risk in patients who are obese and is often used in practice; however, there is little solid evidence to support this. In contrast, aspirin is also a medicine that requires perioperative withdrawal due to bleeding risks. Although withdrawal is often considered unnecessary for minor operations, its disadvantages should be considered in patients undergoing P-FST, who frequently undergo surgeries. Therefore, aspirin is often avoided in combination. When P-FST is administered to obese patients, it is advisable to start it with the risk of blood clots in mind, explaining the risks and recommending weight management, including diet and exercise.

Fertility-sparing treatment is aimed only at achieving pregnancy. However, frequent endometrial examinations and D&C may be necessary for accurate diagnosis and disease follow-up. Intrauterine lesions, such as Asherman syndrome or endometritis, may be noted, but there are no clear reports of an increased risk, specifically in patients with P-FST. Frequent D&C procedures are believed to cause implantation failure, which is essential for pregnancy. However, there are currently no reports indicating a clear increase in implantation failure or infertility in patients following P-FST. Placenta accreta has also been reported as a complication of pregnancy and delivery after P-FST. Details related to pregnancy are provided in the following sections.

11.5 Risk of Synchronous Endometrial and Ovarian Cancers

Ovarian cancer is frequently reported during or after P-FST for endometrial cancer [56–61]. Generally, 3.1–10% of patients with endometrial cancer exhibit ovarian metastasis [62], and among those initially diagnosed as stage I endometrial cancer, ovarian metastasis is observed in approximately 5% [63, 64]. According to Ushijima

[41], 3.6% of patients with endometrial cancer develop ovarian cancer, and 0.24% develop peritoneal cancer. In these cases, most ovarian cancers were endometrioid carcinomas (85.7%), necessitating thorough evaluation to distinguish between ovarian metastases and synchronous primary ovarian cancers. From the perspective of patients with ovarian cancer, it has been reported that 13.3% are metastatic from endometrial cancer, and in most synchronous cases involving endometrial and ovarian cancers, the endometrium is identified as the primary site [65, 66]. Patients undergoing P-FST often undergo D&C procedures more frequently than usual, and if necessary, transcervical resection. It is thought that these procedures may facilitate the dissemination of cancer cells from the cornu of the fallopian tube toward the fallopian tube itself. However, a systematic review on 38 studies concerning hysteroscopy and risk of neoplastic cell dissemination concluded that there is no current evidence that diagnostic hysteroscopy increases the risk of malignant cells spreading into the peritoneal cavity or worsens the prognosis in women with endometrial cancer [67].

Given these risks, clinicians must remain vigilant regarding the potential development of ovarian cancer during or after P-FST. The use of CA125 levels and imaging diagnostics should be considered appropriate along with an evaluation of hereditary cancer syndromes during patient management (see Chap. 12).

11.6 Advanced Case Reports

Fertility-sparing treatment is used only in patients with AEH and early-stage endometrial cancer. As a result, there are few patients with cancer progressing or resulting in death, but it is important to note that there have been reports of extra-uterine metastasis and death during or after P-FST [36, 68]. As mentioned in Sect. 11.5, ovarian metastasis occurs in some cases; however, the disease can also progress to other areas. Careful management is necessary because the standard treatment is hysterectomy. Standard treatments should be adopted immediately when fertility is not preserved. Even after careful consideration of the indications and management, there have been reports of 1.6–2.2% of cases showing extrauterine metastasis [41]. The most common sites of metastasis are regional lymph nodes, peritoneum, and ovaries, but it is said that follow-up imaging and tumor markers are necessary as needed [43, 56–61, 69]. Additionally, there have been reports of cases in which a patient died after a diagnosis of extrauterine metastasis was made and the patient's condition drastically worsened [36]. It is also possible that these cases were genetically high risk, as discussed later; however, careful management should be carried out.

11.7 Prognosis: Infertility Treatment and Pregnancy Outcomes

11.7.1 Infertility Treatment Outcomes

Following P-FST, transitioning to infertility treatment as soon as possible is recommended, owing to the risk of recurrence. Repeated D&C procedures and endometrial thinning, which are often associated with P-FST, have been suggested to negatively affect pregnancy success rates. Additionally, endometrial thickness during ovulation, patient age, and recurrence rate also affected pregnancy success [70]. Inoue et al. showed that increased D&C procedures lead to thinner endometria during ovulation, thereby reducing the likelihood of a successful pregnancy.

However, there have been no reports of patients having particularly poor fertility after P-FST. Pregnancy rates vary, depending on the treatment approach. Reports indicate that infertility treatments, such as assisted reproductive technology (ART), result in pregnancy rates of 66–80% after P-FST [36, 37, 71, 72]. Compared to 43.2% in those who conceived naturally, ART is considered effective for achieving pregnancy after P-FST [72]. Although it is possible to conceive naturally after P-FST, it is important to consider the possibility that it may be difficult to conceive due to the risk of recurrence and the aforementioned risks, and to become pregnant as soon as possible, even with fertility treatment.

11.7.2 Pregnancy Outcomes

The risk of lesion recurrence poses a significant challenge in achieving pregnancy after treatment. Inoue et al. reported a recurrence rate of 62.2% after treatment and found that the higher the age at which pregnancy was attempted, the lower the success rate [70]. Therefore, patients should undergo infertility treatment as early as possible.

Furthermore, studies on the recurrence rates of AEH and endometrial cancer in relation to pregnancy suggest that the risk of recurrence is 65.9% among those who do not achieve pregnancy, whereas it is 26.4% among those who do [37, 41]. Based on these findings, rapid progression to fertility treatment after P-FST with the goal of achieving pregnancy is highly desirable.

11.7.3 Potential Adverse Effects During Pregnancy and Delivery

The potential effects of P-FST on pregnancy and delivery outcomes in AEH and early-stage endometrial cancer have also been examined. Compared to patients who did not undergo the P-FST, no significant differences were observed in the rates of hypertensive disorders of pregnancy, gestational diabetes mellitus, or threatened preterm labor. However, an increased risk of placenta accreta has been reported. Increased risk is thought to be associated with frequent D&C procedures or endometrial defects resulting from P-FST.

However, Oishi et al. assessed placenta accreta based on whether manual separation of the placenta was required rather than on pathological confirmation [73]. In reality, diagnosing placenta accreta based on histopathological examination of the resected uterus is nearly impossible in pregnancies following P-FST. However, the increased risk of requiring manual placental separation alone is not sufficient to mandate cesarean delivery as a mode of childbirth after P-FST. Excluding the risks associated with advanced maternal age, few pregnancy- or delivery-specific risks were directly attributable to P-FST.

11.8 Genomic Perspectives

Traditionally, endometrial cancer has been classified into two types: estrogen-dependent type I and estrogen non-dependent type II. However, with advancements in molecular genetic analysis, the Cancer Genome Atlas (TCGA) project now categorizes endometrial cancer into four subtypes based on comprehensive molecular profiling: polymerase epsilon (POLE) ultramutated, MSI (hypermutated) (MSI-H), copy-number high (serous-like), and copy-number low (endometrioid) [74]. Alternatively, the clinically accessible Proactive Molecular Risk Classifier for Endometrial Cancer (ProMisE classification) could be used. The ProMisE classification utilizes formalin-fixed, paraffin-embedded biopsy samples and categorizes endometrial cancer into four groups: mismatch repair deficient (MMR-D), POLE EDM, p53 abnormal, and p53 wild-type, based on immunohistochemical analyses of MMR and p53, as well as the detection of POLE mutation (See Chap. 7).

Since the International Federation of Gynecology and Obstetrics (FIGO) staging system for endometrial cancer was revised in 2023 to introduce a new stage, significant advances in molecular biology and genome sequencing have led to further updates [75–77]. In early-stage endometrial cancer, POLE mutations and p53 abnormalities influence the FIGO stage. Specifically, cases with POLE mutations are classified as stage $IAm_{POLEmut}$, regardless of myometrial invasion (lymphovascular space invasion) or histological type. Similarly, cases with p53 abnormalities (p53 abn) were classified as stage $IICm_{p53abn}$, regardless of cervical or myometrial invasion. This marks a groundbreaking shift in cancer staging, moving beyond the traditional approach based solely on the extent of cancer spread, to incorporate genomic abnormalities as critical determinants.

11.8.1 Implications of Molecular Subtypes in P-FST of Endometrial Cancer

The role of molecular subtypes in P-FST of endometrial cancer is of increasing interest. Studies have reported that for the MSI-H/MMRd subtype, the CR rate is 61.8%, with a recurrence rate of 41.2%, which contributes to poor outcomes in P-FST. Although some studies indicate no significant difference in recurrence rates compared to the general population, they report lower remission rates [78–80]. Conversely, other studies have suggested that the MSI-H/MMRd status does not significantly affect the outcomes of P-FSTs, highlighting the need for further investigation [78]. Bani et al. performed a prognostic classification in PFST based on the prognostic classification of the ProMisE system (Fig. 11.3) [81]. Based on this, Bani et al. performed a prognostic classification in P-FST [81]. Fertility-sparing treatment is not contraindicated for patients with hereditary diseases or those who

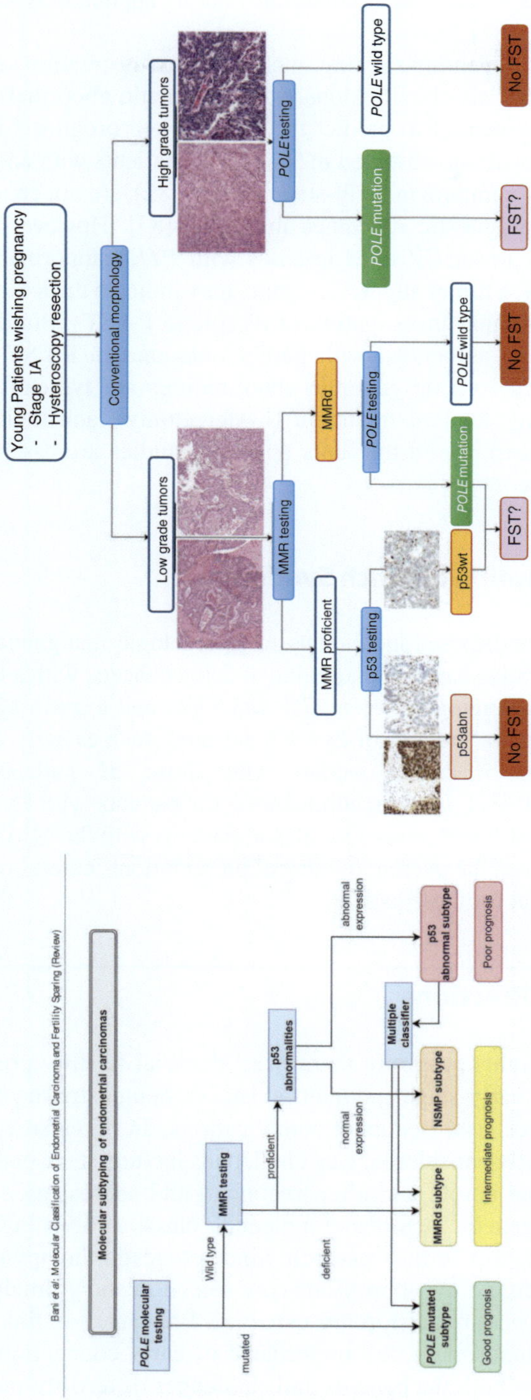

Fig. 11.3 Molecular classification of endometrial cancer and fertility-sparing therapy (FST) eligibility. (Adapted from Bani et al. [81])

may have a predisposition to such conditions, but its applicability should be carefully evaluated [6, 7, 81].

With the advent of genomic testing, including next-generation sequencing and the TCGA and ProMisE classifcations, various genetic abnormalities are being explored for their potential to predict the efficacy and prognosis of P-FST. For example, *PTEN* mutations, observed in 57–83% of patients with endometrial cancer and particularly common in early-stage disease [82], are not consistently associated with poor prognostic outcomes in P-FST [83]. However, other studies report significantly lower CR rates in cases with *PTEN* abnormalities [84, 85]. Mutations in *PTEN*, a tumor suppressor gene, may indicate difficulty in achieving CR, necessitating further investigation of its role in P-FSTs. Similarly, *PIK3CA* mutations have been associated with poorer outcomes in P-FST [85]. Despite these findings, such somatic genomic abnormalities are typically identified in patients undergoing standard treatment (hysterectomy), and their relevance to P-FST remains uncertain. Further data and case studies are essential to clarify their utility in this context.

11.8.2 Consideration of Lynch Syndrome

Lynch syndrome predisposes individuals to gynecologic malignancies, including endometrial and ovarian cancers in addition to colon cancers, with a lifetime risk of endometrial cancer ranging between 27% and 71%, and a mean age of 48 years [86]. The clinical characteristics of Lynch syndrome, such as early onset and concurrent colon cancer, often overlap with those of patients undergoing P-FST. Currently, P-FST is not recommended for patients with Lynch syndrome because of reports of lower remission rates in patients with the MMRd subtype [6, 7]. Nonetheless, given the overlap in clinical presentations, careful monitoring and follow-up are crucial for these patients.

11.9 Future Directions

This chapter highlights progestin therapy as a crucial fertility-preserving treatment for AEH and early-stage endometrial cancer, demonstrating high CR rates. However, recurrence risk, treatment complications, and pregnancy success factors must be carefully considered. Key challenges include strict patient selection, careful management before and after treatment, and addressing obesity-related risks. The role of genetic factors and molecular classifications in CR and recurrence rates also requires further research. Since progestin therapy is a temporary alternative to hysterectomy, appropriate case selection and a timely transition to fertility treatment post-remission are essential. Pregnancy is linked to a lower recurrence risk, emphasizing the importance of early conception efforts. This chapter underscores both the benefits and challenges of fertility-preserving therapy, providing a foundation for future research and clinical practice to optimize patient outcomes.

References

1. WHO cancer today. https://gco.iarc.fr/today/en/dataviz/bars?mode=cancer&types=0&group_populations=1&sort_by=value0&populations=900&sexes=2&key=total&age_end=14.
2. Ansbro É, Issa R, Willis R, Blanchet K, Perel P, Roberts B. Chronic NCD care in crises: a qualitative study of global experts' perspectives on models of care for hypertension and diabetes in humanitarian settings. J Migr Health. 2022;5:100094. https://doi.org/10.1016/j.jmh.2022.100094.
3. Minig L, Franchi D, Boveri S, Casadio C, Bocciolone L, Sideri M, et al. Ann Oncol. 2011;22(3):643–9. https://doi.org/10.1093/annonc/mdq463.
4. Liu Q, Zhou H, Yu M, Cao D, Yang J. GnRH-a-based fertility-sparing treatment of atypical endometrial hyperplasia (AEH) and early endometrial carcinoma (EC) patients: a multicenter, open-label, randomized designed clinical trial protocol. Trials. 2024;25(1):578. https://doi.org/10.1186/s13063-024-08414-0.
5. Cockrum RH, Soo J, Ham SA, Cohen KS, Snow SG. Association of Progestogens and Venous Thromboembolism among women of reproductive age. Obstet Gynecol. 2022;140(3):477–87. https://doi.org/10.1097/AOG.0000000000004896.
6. Concin N, Matias-Guiu X, Vergote I, Cibula D, Mirza MR, Marnitz S, Ledermann J, Bosse T, Chargari C, Fagotti A, Fotopoulou C, Gonzalez Martin A, Lax S, Lorusso D, Marth C, Morice P, Nout RA, O'Donnell D, Querleu D, Raspollini MR, Sehouli J, Sturdza A, Taylor A, Westermann A, Wimberger P, Colombo N, Planchamp F, Creutzberg CL. ESGO/ESTRO/ESP guidelines for the management of patients with endometrial carcinoma. Int J Gynecol Cancer. 2021;31(1):12–39. https://doi.org/10.1136/ijgc-2020-002230.
7. Rodolakis A, Scambia G, Planchamp F, Acien M, Di Spiezio Sardo A, Farrugia M, Grynberg M, Pakiz M, Pavlakis K, Vermeulen N, Zannoni G, Zapardiel I, KLT M. ESGO/ESHRE/ESGE guidelines for the fertility-sparing treatment of patients with endometrial carcinoma. Hum Reprod Open. 2023;2023(1):hoac057. https://doi.org/10.1093/hropen/hoac057.
8. Abu-Rustum N, Yashar C, Arend R, Barber E, Bradley K, Brooks R, Campos SM, Chino J, Chon HS, Chu C, Crispens MA, Damast S, Fisher CM, Frederick P, Gaffney DK, Giuntoli R, Han E, Holmes J, Howitt BE, Lea J, Mariani A, Mutch D, Nagel C, Nekhlyudov L, Podoll M, Salani R, Schorge J, Siedel J, Sisodia R, Soliman P, Ueda S, Urban R, Wethington SL, Wyse E, Zanotti K, NR MM, Aggarwal S. Uterine neoplasms, version 1.2023, NCCN clinical practice guidelines in oncology. J Natl Compr Cancer Netw. 2023;21(2):181–209. https://doi.org/10.6004/jnccn.2023.0006.
9. Piatek S, Michalski W, Sobiczewski P, Bidzinski M, Szewczyk G. The results of different fertility-sparing treatment modalities and obstetric outcomes in patients with early endometrial cancer and atypical endometrial hyperplasia: case series of 30 patients and systematic review. Eur J Obstet Gynecol Reprod Biol. 2021;263:139–47. https://doi.org/10.1016/j.ejogrb.2021.06.007.
10. Gunderson CC, Fader AN, Carson KA, Bristow RE. Oncologic and reproductive outcomes with progestin therapy in women with endometrial hyperplasia and grade 1 adenocarcinoma: a systematic review. Gynecol Oncol. 2012;125(2):477–82. https://doi.org/10.1016/j.ygyno.2012.01.003.
11. Lucchini SM, Esteban A, Nigra MA, Palacios AT, Alzate-Granados JP, Borla HF. Updates on conservative management of endometrial cancer in patients younger than 45 years. Gynecol Oncol. 2021;161(3):802–9. https://doi.org/10.1016/j.ygyno.2021.04.017.
12. Xu Z, Yang B, Shan W, Liao J, Shao W, Wu P, Zhou S, Ning C, Luo X, Zhu Q, Zhang H, Ma F, Guan J, Chen X. Comparison of the effect of levonorgestrel-intrauterine system with or without oral megestrol acetate on fertility-preserving treatment in patients with atypical endometrial hyperplasia: a prospective, open-label, randomized controlled phase II study. Gynecol Oncol. 2023;174:133–41. https://doi.org/10.1016/j.ygyno.2023.05.001.

13. Luo L, Luo B, Zheng Y, Zhang H, Li J, Sidell N. Oral and intrauterine progestogens for atypical endometrial hyperplasia. Cochrane Database Syst Rev. 2018;12(12):CD009458. https://doi.org/10.1002/14651858.CD009458.pub3.
14. Kim MK, Seong SJ, Song T, Kim ML, Yoon BS, Jun HS, Lee GH, Lee YH. Comparison of dilatation & curettage and endometrial aspiration biopsy accuracy in patients treated with high-dose oral progestin plus levonorgestrel intrauterine system for early-stage endometrial cancer. Gynecol Oncol. 2013;130(3):470–3. https://doi.org/10.1016/j.ygyno.2013.06.035.
15. Kim MK, Seong SJ, Lee TS, Kim JW, Nam BH, Hong SR, et al. Treatment with medroxyprogesterone acetate plus levonorgestrel-releasing intrauterine system for early-stage endometrial cancer in young women: single-arm, prospective multicenter study: Korean Gynecologic Oncology Group Study (KGOG2009). Jpn J Clin Oncol. 2012;42(12):1215–8.
16. Kim MK, Seong SJ, Kim YS, Song T, Kim ML, Yoon BS, et al. Combined medroxyprogesterone acetate/levonorgestrel–intrauterine system treatment in young women with early-stage endometrial cancer. Am J Obstet Gynecol. 2013;209(4):358.e1–4.
17. Kim MK, Seong SJ, Kang SB, Bae DS, Kim JW, Nam JH, et al. Six months response rate of combined oral medroxyprogesterone/levonorgestrel-intrauterine system for early-stage endometrial cancer in young women: a Korean Gynecologic-Oncology Group Study. J Gynecol Oncol. 2019;30(2):e47.
18. Kim DH, Seong SJ, Kim MK, Bae HS, Kim M, Yun BS, Jung YW, Shim JY. Dilatation and curettage is more accurate than endometrial aspiration biopsy in early-stage endometrial cancer patients treated with high dose oral progestin and levonorgestrel intrauterine system. J Gynecol Oncol. 2017;28(1):e1. https://doi.org/10.3802/jgo.2017.28.e1.
19. Gu W, Mitsuhashi A, Kobayashi T, Shozu M. Metformin attenuates the production and proliferative effects of prolactin induced by medroxyprogesterone acetate during fertility-sparing treatment for endometrial cancer. BMC Cancer. 2022;22(1):753. https://doi.org/10.1186/s12885-022-09858-w.
20. Mitsuhashi A, Sato Y, Kiyokawa T, Koshizaka M, Hanaoka H, Shozu M. Phase II study of medroxyprogesterone acetate plus metformin as a fertility-sparing treatment for atypical endometrial hyperplasia and endometrial cancer. Ann Oncol. 2016;27(2):262–6. https://doi.org/10.1093/annonc/mdv539.
21. Zhou H, Cao D, Yang J, Shen K, Lang J. Gonadotropin-releasing hormone agonist combined with a levonorgestrel-releasing intrauterine system or letrozole for fertility-preserving treatment of endometrial carcinoma and complex atypical hyperplasia in young women. Int J Gynecol Cancer. 2017;27(6):1178–82. https://doi.org/10.1097/IGC.0000000000001008.
22. Mitsuhashi A, Kawasaki Y, Hori M, Fujiwara T, Hanaoka H, Shozu M. Medroxyprogesterone acetate plus metformin for fertility-sparing treatment of atypical endometrial hyperplasia and endometrial carcinoma: trial protocol for a prospective, randomised, open, blinded-endpoint design, dose-response trial (FELICIA trial). BMJ Open. 2020;10(2):e035416. https://doi.org/10.1136/bmjopen-2019-035416.
23. Rothwell PM, Howard SC, Dolan E, O'Brien E, Dobson JE, Dahlöf B, Sever PS, Poulter NR. Prognostic significance of visit-to-visit variability, maximum systolic blood pressure, and episodic hypertension. Lancet. 2010;375(9718):895–905. https://doi.org/10.1016/S0140-6736(10)60308-X.
24. Cook NR, Lee IM, Zhang SM, Moorthy MV, Buring JE. Alternate-day, low-dose aspirin and cancer risk: long-term observational follow-up of a randomized trial. Ann Intern Med. 2013;159(2):77–85. https://doi.org/10.7326/0003-4819-159-2-201307160-00002.
25. Takiuchi T, Blake EA, Matsuo K, Sood AK, Brasky TM. Aspirin use and endometrial cancer risk and survival. Gynecol Oncol. 2018;148(1):222–32. https://doi.org/10.1016/j.ygyno.2017.10.026.
26. Chang X, Liu SJ, Han L. Causal association between aspirin use and risk of endometrioid carcinoma: a Mendelian randomization study. Eur Rev Med Pharmacol Sci. 2023;27(23):11597–605. https://doi.org/10.26355/eurrev_202312_34598.

27. JJ MN, Wolfe R, Woods RL, Tonkin AM, Donnan GA, Nelson MR, Reid CM, Lockery JE, Kirpach B, Storey E, Shah RC, Williamson JD, Margolis KL, Ernst ME, Abhayaratna WP, Stocks N, Fitzgerald SM, Orchard SG, Trevaks RE, Beilin LJ, Johnston CI, Ryan J, Radziszewska B, Jelinek M, Malik M, Eaton CB, Brauer D, Cloud G, Wood EM, Mahady SE, Satterfield S, Grimm R, Murray AM, ASPREE Investigator Group. Effect of aspirin on cardiovascular events and bleeding in the healthy elderly. N Engl J Med. 2018;379(16):1509–18. https://doi.org/10.1056/NEJMoa1805819.
28. Yoshimura T, Yamagami W, Takahashi M, Hirano T, Sakai K, Makabe T, et al. Clinical usefulness of endometrial cytology in determining the therapeutic effect of fertility preserving therapy. Acta Cytol. 2022;66(2):106–13.
29. Carugno J, Wong A. Fertility-sparing approach for endometrial cancer: the role of office hysteroscopy. Minim Invasive Ther Allied Technol. 2021;30(5):296–303. https://doi.org/10.1080/13645706.2021.1949353.
30. Park JY. Hysteroscopy in fertility-sparing management for early endometrial cancer: a double-edged sword. J Gynecol Oncol. 2017;28(1):e16. https://doi.org/10.3802/jgo.2017.28.e16.
31. Bilir E, Kahramanoğlu İ. The role of hysteroscopy in fertility preservation in endometrial cancer and atypical endometrial hyperplasia: a semi-systematic literature review. Arch Gynecol Obstet. 2023;308(4):1113–26. https://doi.org/10.1007/s00404-023-06960-7.
32. Guillon S, Popescu N, Phelippeau J, Koskas M. A systematic review and meta-analysis of prognostic factors for remission in fertility-sparing management of endometrial atypical hyperplasia and adenocarcinoma. Int J Gynaecol Obstet. 2019;146(3):277–88. https://doi.org/10.1002/ijgo.12882.
33. Giampaolino P, Di Spiezio Sardo A, Mollo A, Raffone A, Travaglino A, Boccellino A, Zizolfi B, Insabato L, Zullo F, De Placido G, Bifulco G. Hysteroscopic endometrial focal resection followed by levonorgestrel intrauterine device insertion as a fertility-sparing treatment of atypical endometrial hyperplasia and early endometrial cancer: a retrospective study. J Minim Invasive Gynecol. 2019;26(4):648–56. https://doi.org/10.1016/j.jmig.2018.07.001.
34. Dong H, Wang Y, Zhang M, Sun M, Yue Y. Whether preoperative hysteroscopy increases the dissemination of endometrial cancer cells: a systematic review and meta-analysis. J Obstet Gynaecol Res. 2021;47(9):2969–77. https://doi.org/10.1111/jog.14897.
35. Aumiphin J, Crochet P, Knight S, Carcopino X, Cravello L, Boubli L, Agostini A. Outcome and follow-up of patients with endometrial carcinoma diagnosed on operative hysteroscopic resection specimens. Anticancer Res. 2016;36(8):4341–5.
36. Ushijima K, Yahata H, Yoshikawa H, Konishi I, Yasugi T, Saito T, Nakanishi T, Sasaki H, Saji F, Iwasaka T, Hatae M, Kodama S, Saito T, Terakawa N, Yaegashi N, Hiura M, Sakamoto A, Tsuda H, Fukunaga M, Kamura T. Multicenter phase II study of fertility-sparing treatment with medroxyprogesterone acetate for endometrial carcinoma and atypical hyperplasia in young women. J Clin Oncol. 2007;25(19):2798–803. https://doi.org/10.1200/JCO.2006.08.8344.
37. Yamagami W, Susumu N, Makabe T, Sakai K, Nomura H, Kataoka F, Hirasawa A, Banno K, Aoki D. Is repeated high-dose medroxyprogesterone acetate (MPA) therapy permissible for patients with early stage endometrial cancer or atypical endometrial hyperplasia who desire preserving fertility? J Gynecol Oncol. 2018;29(2):e21. https://doi.org/10.3802/jgo.2018.29.e21.
38. Ohyagi-Hara C, Sawada K, Aki I, Mabuchi S, Kobayashi E, Ueda Y, Yoshino K, Fujita M, Tsutsui T, Kimura T. Efficacies and pregnant outcomes of fertility-sparing treatment with medroxyprogesterone acetate for endometrioid adenocarcinoma and complex atypical hyperplasia: our experience and a review of the literature. Arch Gynecol Obstet. 2015;291(1):151–7. https://doi.org/10.1007/s00404-014-3417-z.
39. Fujiwara H, Jobo T, Takei Y, Saga Y, Imai M, Arai T, Taneichi A, Machida S, Takahashi Y, Suzuki M. Fertility-sparing treatment using medroxyprogesterone acetate for endometrial carcinoma. Oncol Lett. 2012;3(5):1002–6. https://doi.org/10.3892/ol.2012.602.
40. Park JY, Kim DY, Kim JH, Kim YM, Kim KR, Kim YT, Seong SJ, Kim TJ, Kim JW, Kim SM, Bae DS, Nam JH. Long-term oncologic outcomes after fertility-sparing management using oral progestin for young women with endometrial cancer (KGOG 2002). Eur J Cancer. 2013;49(4):868–74. https://doi.org/10.1016/j.ejca.2012.09.017.

41. Ushijima K, Tsuda N, Yamagami W, Mitsuhashi A, Mikami M, Yaegashi N, Enomoto T. Trends and characteristics of fertility-sparing treatment for atypical endometrial hyperplasia and endometrial cancer in Japan: a survey by the Gynecologic Oncology Committee of the Japan Society of obstetrics and Gynecology. J Gynecol Oncol. 2023;34(3):e38. https://doi.org/10.3802/jgo.2023.34.e38.
42. Gallos ID, Yap J, Rajkhowa M, Luesley DM, Coomarasamy A, Gupta JK. Regression, relapse, and live birth rates with fertility-sparing therapy for endometrial cancer and atypical complex endometrial hyperplasia: a systematic review and metaanalysis. Am J Obstet Gynecol. 2012;207(4):266.e1–12. https://doi.org/10.1016/j.ajog.2012.08.011.
43. Kaku T, Yoshikawa H, Tsuda H, Sakamoto A, Fukunaga M, Kuwabara Y, Hataeg M, Kodama S, Kuzuya K, Sato S, Nishimura T, Hiura M, Nakano H, Iwasaka T, Miyazaki K, Kamura T. Conservative therapy for adenocarcinoma and atypical endometrial hyperplasia of the endometrium in young women: central pathologic review and treatment outcome. Cancer Lett. 2001;167(1):39–48. https://doi.org/10.1016/s0304-3835(01)00462-1.
44. Fan Z, Li H, Hu R, Liu Y, Liu X, Gu L. Fertility-preserving treatment in young women with grade 1 presumed stage IA endometrial adenocarcinoma: a meta-analysis. Int J Gynecol Cancer. 2018;28(2):385–93. https://doi.org/10.1097/IGC.0000000000001164.
45. Zhao S, Zhang J, Yan Y, Tian L, Chen L, Zheng X, Sun Y, Tian W, Xue F, Wang Y. Oncological and reproductive outcomes of endometrial atypical hyperplasia and endometrial cancer patients undergoing conservative therapy with hysteroscopic resection: a systematic review and meta-analysis. Acta Obstet Gynecol Scand. 2024;103(8):1498–512. https://doi.org/10.1111/aogs.14815.
46. Herrera Cappelletti E, Humann J, Torrejón R, Gambadauro P. Chances of pregnancy and live birth among women undergoing conservative management of early-stage endometrial cancer: a systematic review and meta-analysis. Hum Reprod Update. 2022;28(2):282–95. https://doi.org/10.1093/humupd/dmab041.
47. Wei J, Zhang W, Feng L, Gao W. Comparison of fertility-sparing treatments in patients with early endometrial cancer and atypical complex hyperplasia: a meta-analysis and systematic review. Medicine (Baltimore). 2017;96(37):e8034. https://doi.org/10.1097/MD.0000000000008034.
48. Murakami I, Machida H, Morisada T, Terao Y, Tabata T, Mikami M, Hirashima Y, Kobayashi Y, Baba T, Nagase S. Effects of a fertility-sparing re-treatment for recurrent atypical endometrial hyperplasia and endometrial cancer: a systematic literature review. J Gynecol Oncol. 2023;34(4):e49. https://doi.org/10.3802/jgo.2023.34.e49.
49. Chen J, Cao D. Fertility-sparing re-treatment for endometrial cancer and atypical endometrial hyperplasia patients with progestin-resistance: a retrospective analysis of 61 cases. World J Surg Oncol. 2024;22(1):169. https://doi.org/10.1186/s12957-024-03439-w.
50. Sakai K, Yamagami W, Sato Y, Susumu N, Yokoyama Y, Takehara K, Mandai M, Okamoto A. A phase II trial evaluating the efficacy and safety of repeated high dose medroxyprogesterone acetate (MPA) therapy for patients with recurrent early-stage endometrial cancer or atypical endometrial hyperplasia: Japanese Gynecologic Oncology Group study (JGOG2051/KGOG2031, REMPA trial). J Gynecol Oncol. 2024;35(6):e106. https://doi.org/10.3802/jgo.2024.35.e106.
51. Giampaolino P, Cafasso V, Boccia D, Ascione M, Mercorio A, Viciglione F, Palumbo M, Serafino P, Buonfantino C, De Angelis MC, Verrazzo P, Grasso G, Gullo G, Bifulco G, Della Corte L. Fertility-sparing approach in patients with endometrioid endometrial cancer grade 2 stage IA (FIGO): a qualitative systematic review. Biomed Res Int. 2022;2022:4070368. https://doi.org/10.1155/2022/4070368.
52. Lee AJ, Yang EJ, Kim NK, Kim Y, Suh DH, Kim J, Son JH, Kong TW, Chang SJ, Hwang DW, Park SJ, Kim HS, Yoo JG, Lee SJ, Lee YY, Shim SH. Fertility-sparing hormonal treatment in patients with stage I endometrial cancer of grade 2 without myometrial invasion and grade 1-2 with superficial myometrial invasion: Gynecologic oncology research investigators coLLaborAtion study (GORILLA-2001). Gynecol Oncol. 2023;174:106–13. https://doi.org/10.1016/j.ygyno.2023.04.027.

53. Yu M, Wang Y, Yuan Z, Zong X, Huo X, Cao DY, Yang JX, Shen K. Fertility-sparing treatment in young patients with grade 2 presumed stage IA endometrioid endometrial adenocarcinoma. Front Oncol. 2020;10:1437. https://doi.org/10.3389/fonc.2020.01437.
54. Rosano GMC, Rodriguez-Martinez MA, Spoletini I, Regidor PA. Obesity and contraceptive use: impact on cardiovascular risk. ESC Heart Fail. 2022;9(6):3761–7. https://doi.org/10.1002/ehf2.14104. Epub 2022 Sep 14. Erratum in: ESC Heart Fail. 2023;10(2):1497
55. Stein PD, Beemath A, Olson RE. Obesity as a risk factor in venous thromboembolism. Am J Med. 2005;118(9):978–80. https://doi.org/10.1016/j.amjmed.2005.03.012.
56. Randall TC, Kurman RJ. Progestin treatment of atypical hyperplasia and well-differentiated carcinoma of the endometrium in women under age 40. Obstet Gynecol. 1997;90(3):434–40. https://doi.org/10.1016/s0029-7844(97)00297-4.
57. Duska LR, Garrett A, Rueda BR, Haas J, Chang Y, Fuller AF. Endometrial cancer in women 40 years old or younger. Gynecol Oncol. 2001;83(2):388–93. https://doi.org/10.1006/gyno.2001.6434.
58. Pinto AB, Gopal M, Herzog TJ, Pfeifer JD, Williams DB. Successful in vitro fertilization pregnancy after conservative management of endometrial cancer. Fertil Steril. 2001;76(4):826–9. https://doi.org/10.1016/s0015-0282(01)01983-5.
59. Corrado G, Baiocco E, Carosi M, Vizza E. Progression of conservatively treated endometrial complex atypical hyperplasia in a young woman: a case report. Fertil Steril. 2008;90(5):2006.e5–8. https://doi.org/10.1016/j.fertnstert.2008.06.014.
60. Huang SY, Jung SM, Ng KK, Chang YC, Lai CH. Ovarian metastasis in a nulliparous woman with endometrial adenocarcinoma failing conservative hormonal treatment. Gynecol Oncol. 2005;97(2):652–5. https://doi.org/10.1016/j.ygyno.2005.01.035.
61. Takeshima N, Hirai Y, Yano K, Tanaka N, Yamauchi K, Hasumi K. Ovarian metastasis in endometrial carcinoma. Gynecol Oncol. 1998;70(2):183–7. https://doi.org/10.1006/gyno.1998.5086.
62. Al Hilli MM, Dowdy SC, Weaver AL, St Sauver JL, Keeney GL, Mariani A, Podratz KC, Bakkum-Gamez JN. Incidence and factors associated with synchronous ovarian and endometrial cancer: a population-based case-control study. Gynecol Oncol. 2012;125(1):109–13. https://doi.org/10.1016/j.ygyno.2011.12.444.
63. Creasman WT, DeGeest K, DiSaia PJ, Zaino RJ. Significance of true surgical pathologic staging: a Gynecologic oncology group study. Am J Obstet Gynecol. 1999;181(1):31–4. https://doi.org/10.1016/s0002-9378(99)70431-x.
64. Creasman WT, Morrow CP, Bundy BN, Homesley HD, Graham JE, Heller PB. Surgical pathologic spread patterns of endometrial cancer. A Gynecologic oncology group study. Cancer. 1987;60(8 Suppl):2035–41. https://doi.org/10.1002/1097-0142(19901015)60:8+<2035::aid-cncr2820601515>3.0.co;2-8.
65. Sakamoto I, Hirotsu Y, Amemiya K, Nozaki T, Mochizuki H, Omata M. Elucidation of genomic origin of synchronous endometrial and ovarian cancer (SEO) by genomic and microsatellite analysis. J Gynecol Oncol. 2023;34(1):e6. https://doi.org/10.3802/jgo.2023.34.e6.
66. Chao A, Wu RC, Jung SM, Lee YS, Chen SJ, Lu YL, Tsai CL, Lin CY, Tang YH, Chen MY, Huang HJ, Chou HH, Huang KG, Chang TC, Wang TH, Lai CH. Implication of genomic characterization in synchronous endometrial and ovarian cancers of endometrioid histology. Gynecol Oncol. 2016;143(1):60–7. https://doi.org/10.1016/j.ygyno.2016.07.114.
67. Yazbeck C, Dhainaut C, Batallan A, Benfila J, Thoury A, Madelenat P. Hysteroscopie diagnostique et risque de contamination peritoneale par les cellules tumorales. Gynecol Obstet Fertil. 2005;33:247–52.
68. Gitsch G, Hanzal E, Jensen D, Hacker NF. Endometrial cancer in premenopausal women 45 years and younger. Obstet Gynecol. 1995;85(4):504–8. https://doi.org/10.1016/0029-7844(95)00001-8.
69. Rose PG, Mendelsohn G, Kornbluth I. Hysteroscopic dissemination of endometrial carcinoma. Gynecol Oncol. 1998;71(1):145–6. https://doi.org/10.1006/gyno.1998.5139.

70. Inoue O, Hamatani T, Susumu N, Yamagami W, Ogawa S, Takemoto T, Hirasawa A, Banno K, Kuji N, Tanaka M, Aoki D. Factors affecting pregnancy outcomes in young women treated with fertility-preserving therapy for well-differentiated endometrial cancer or atypical endometrial hyperplasia. Reprod Biol Endocrinol. 2016;14:2. https://doi.org/10.1186/s12958-015-0136-7.
71. Vasileva R, Wohrer H, Gaultier V, Bucau M, Courcier H, Ben Miled S, Gonthier C, Koskas M. Pregnancy and obstetric outcomes after fertility-sparing management of endometrial cancer and atypical hyperplasia: a multicentre cohort study. Hum Reprod. 2024;39(6):1231–8. https://doi.org/10.1093/humrep/deae089.
72. Tong XM, Lin XN, Jiang HF, Jiang LY, Zhang SY, Liang FB. Fertility-preserving treatment and pregnancy outcomes in the early stage of endometrial carcinoma. Chin Med J. 2013;126(15):2965–71.
73. Oishi M, Kasuga Y, Fukuma Y, Hamuro A, Tamai J, Tanaka Y, Hasegawa K, Yoshimura T, Ikenoue S, Ochiai D, Yamagami W, Tanaka M. Obstetric outcomes after medroxyprogesterone acetate treatment for early stage endometrial cancer or atypical endometrial hyperplasia: a single hospital-based study. Int J Clin Oncol. 2023;28(4):587–91. https://doi.org/10.1007/s10147-023-02297-y.
74. Cancer Genome Atlas Research Network, Kandoth C, Schultz N, Cherniack AD, Akbani R, Liu Y, Shen H, Robertson AG, Pashtan I, Shen R, Benz CC, Yau C, Laird PW, Ding L, Zhang W, Mills GB, Kucherlapati R, Mardis ER, Levine DA. Integrated genomic characterization of endometrial carcinoma. Nature. 2013;497(7447):67–73. https://doi.org/10.1038/nature12113. Erratum in: Nature. 2013;500(7461):242
75. Falcone F, Normanno N, Losito NS, Scognamiglio G, Esposito Abate R, Chicchinelli N, et al. Application of the proactive molecular risk classifier for endometrial cancer (ProMisE) to patients conservatively treated: outcomes from an institutional series. Eur J Obstet Gynecol Reprod Biol. 2019;240:220–5.
76. Kobayashi-Kato M, Fujii E, Asami Y, Ahiko Y, Hiranuma K, Terao Y, Matsumoto K, Ishikawa M, Kohno T, Kato T, Shiraishi K, Yoshida H. Utility of the revised FIGO2023 staging with molecular classification in endometrial cancer. Gynecol Oncol. 2023;178:36–43. https://doi.org/10.1016/j.ygyno.2023.09.011.
77. Mitric C, Bernardini MQ. Endometrial cancer: transitioning from histology to genomics. Curr Oncol. 2022;29(2):741–57.
78. Zhang T, Zhang X, Peng P, Yang J. Fertility-sparing treatment in MSI-H/MMRd endometrial carcinoma or atypical endometrial hyperplasia: a systematic review and meta-analysis. Eur J Obstet Gynecol Reprod Biol. 2024;302:177–83. https://doi.org/10.1016/j.ejogrb.2024.09.006.
79. Chung YS, Woo HY, Lee JY, Park E, Nam EJ, Kim S, Kim SW, Kim YT. Mismatch repair status influences response to fertility-sparing treatment of endometrial cancer. Am J Obstet Gynecol. 2021;224(4):370.e1–370.e13. https://doi.org/10.1016/j.ajog.2020.10.003.
80. Zakhour M, Cohen JG, Gibson A, Walts AE, Karimian B, Baltayan A, Aoyama C, Garcia L, Dhaliwal SK, Elashoff D, Amneus M, Walsh C. Abnormal mismatch repair and other clinicopathologic predictors of poor response to progestin treatment in young women with endometrial complex atypical hyperplasia and well-differentiated endometrial adenocarcinoma: a consecutive case series. BJOG. 2017;124(10):1576–83. https://doi.org/10.1111/1471-0528.14491.
81. Bani MA, Maulard A, Morice P, Chargari C, Genestie C. Integration of the molecular classification of endometrial carcinoma to select patients for fertility sparing strategies. Anticancer Res. 2024;44(2):445–52.
82. Kommoss S, MK MC, Kommoss F, Leung S, Bunz A, Magrill J, Britton H, Kommoss F, Grevenkamp F, Karnezis A, Yang W, Lum A, Krämer B, Taran F, Staebler A, Lax S, Brucker SY, Huntsman DG, Gilks CB, JN MA, Talhouk A. Final validation of the ProMisE

molecular classifier for endometrial carcinoma in a large population-based case series. Ann Oncol. 2018;29(5):1180–8. https://doi.org/10.1093/annonc/mdy058.

83. Travaglino A, Raffone A, Saccone G, Insabato L, Mollo A, De Placido G, et al. PTEN as a predictive marker of response to conservative treatment in endometrial hyperplasia and early endometrial cancer. A systematic review and meta-analysis. Eur J Obstet Gynecol Reprod Biol. 2018;231:104–10.
84. Xue Y, Dong Y, Lou Y, Lv Q, Shan W, Wang C, et al. PTEN mutation predicts unfavorable fertility preserving treatment outcome in the young patients with endometrioid endometrial cancer and atypical hyperplasia. J Gynecol Oncol. 2023;34(4):e53.
85. Xu Y, Zhao M, Zhang L, Wang T, Wang B, Xue Y, et al. Outcomes of fertility preservation treatments in patients with endometrial cancer with different molecular classifications based on an NGS panel. Front Oncol. 2023;13:1282356.
86. Koornstra JJ, Mourits MJ, Sijmons RH, Leliveld AM, Hollema H, Kleibeuker JH. Management of extracolonic tumours in patients with lynch syndrome. Lancet Oncol. 2009;10(4):400–8. https://doi.org/10.1016/S1470-2045(09)70041-5.

Uterine Sarcoma

12

Hiroshi Asano

Abstract

Soft tissue sarcomas (STSs) are rare, making up less than 1% of all cancers, with uterine sarcomas accounting for 4–9% of uterine malignancies. The most common type is uterine leiomyosarcoma (uLMS), an aggressive cancer that develops from the smooth muscle of the uterus. The primary treatment for uLMS is surgical resection, usually through total abdominal hysterectomy and bilateral adnexectomy. Adjuvant chemotherapy has not shown consistent benefits in early-stage cases, but it is commonly used for advanced or recurrent cases with unresectable lesions. Doxorubicin (Dox) is the standard first-line chemotherapy, though combining it with trabectedin has shown better survival rates in clinical trials. Genetic studies have identified mutations in uLMS, including *TP53*, *RB1*, and *BRCA2*. New treatments, such as poly adenosine diphosphate-ribose polymerase (PARP) inhibitors for HRD (homologous recombination deficiency)-positive cases, are being explored. Since uLMS is characterized by chromosomal instability, future therapies may need to be tailored based on genetic and chromosomal alterations. Due to its rarity, innovative strategies, such as large-scale registry studies and genomic data collection, are needed to evaluate treatment effectiveness and develop better therapies.

Keywords

Uterine leiomyosarcomas · Aneuploidy · Homologous recombination deficiency

H. Asano (✉)
Department of Obstetrics, Hokkaido University Hospital, Sapporo, Japan
e-mail: asano.hj@pop.med.hokudai.ac.jp

H. Watari, T. Mitamura (eds.), *Uterine Body Neoplasms*, Comprehensive Gynecology and Obstetrics, https://doi.org/10.1007/978-981-95-0946-1_12

12.1 Introduction

Soft tissue sarcoma (STS) is an extremely rare malignancy of the mesenchymal tissues. STS is derived from the mesoderm and can originate from various tissues anywhere in the body, such as fibrous tissue, adipose tissue, muscle tissue, and vascular tissue. Thus, the histologic subtypes are highly diverse and classified into over 50 types. The incidence of STS is less than 1% of all malignancies [1]. Data from the American Cancer Society shows that 13,520 cases of STS occur annually and 5410 deaths in the US [2]. The prognosis for STS varies greatly depending on factors such as the specific subtype, tumor size, grade, and stage at diagnosis. Treatment typically involves a multidisciplinary approach, often combining surgery, radiation therapy, and chemotherapy. Due to the rarity and complexity of these tumors, the management of STS patients is usually complex.

STS occurring in the uterus is called uterine sarcoma, and it has been reported to account for 4–9% of uterine malignancies. However, since the fifth edition of the WHO classification, carcinosarcomas, which account for 5% of uterine malignancies, have been classified as epithelial neoplasms, making uterine sarcomas even rarer malignancies [3]. Uterine sarcomas are mainly classified as uterine leiomyosarcomas (uLMS), low-grade endometrial stromal sarcomas (LG-ESS), high-grade ESS (HG-ESS), and undifferentiated uterine sarcomas (UUS). As shown in Fig. 12.1, the proportion of each histological subtype of uterine sarcoma is highest for uLMS at 55–60%, followed by ESS at 30–35% and other histological subtypes at 10%, including UUS and an even rarer subtype such as rhabdomyosarcomas.

Uterine sarcomas have conventionally been classified based on their morphological features, incorporating cytologic atypia, tumor cell necrosis, and mitotic rate. However, advances in molecular genetics have significantly improved the ability to differentiate histological subtypes and identify novel treatment targets. Specific translocations involving PGR and PLAG1 have been observed in epithelioid and myxoid LMS [4]. Additionally, a newly recognized subgroup includes

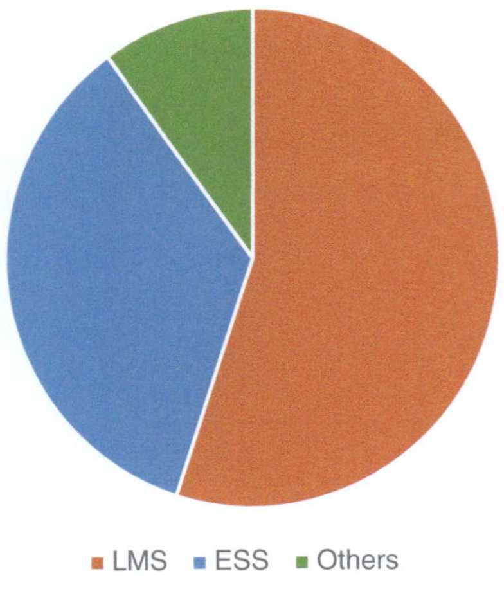

Fig. 12.1 Proportion of histological subtype of uterine sarcoma. LMS leiomyosarcoma, ESS endometrial stromal sarcomas

tumors with ZC3H7B-BCOR fusion and BCOR internal tandem duplication (ITD) in ESS, which share morphological features with myxoid LMS. Meanwhile, with the integration of molecular data and morphological analysis, undifferentiated uterine sarcoma may soon be phased out [3].

Furthermore, the treatment strategy and prognosis of uterine sarcomas vary greatly depending on the histological subtype. The 5-year overall survival (5-y OS) rate for uLMS is abysmal, but LG-ESS is relatively good, at over 90% for stages I and II and appropriately 50% for stages III and IV. However, they have various histological features, so diagnosis can often be difficult even when the same histological type is present. In developing novel treatment strategies, clinically validated next-generation sequencing (NGS) could provide new insights for identifying subsets of patients whose tumors harbor targetable mutations [5].

In this chapter, we focus on uLMS, the most common subtype of uterine sarcoma, and provide an overview of current treatment strategies. We will also discuss the characteristics of genomic abnormalities revealed in recent years and the possibility of future treatment development.

12.2 Clinical Features of uLMS

uLMS is defined as a malignant mesenchymal tumor of myometrial smooth muscle derivation exhibiting spindle cell, epithelioid, or myxoid morphology and divided into three subtypes: spindle, epithelioid, and myxoid LMS [3]. Tumor cells are often spindle-shaped or show marked nuclear pleomorphism, and multinucleated cells are frequently seen but may also show uniform nuclear findings. More than four mitoses per mm^2 (more than 10 in 10 high-power fields) are seen. Tumors are differentiated from leiomyomas by the coagulation necrosis of the tumor and the appearance of infiltration at the tumor border. A universal classification for evaluating the degree of atypia has not been established. In epithelioid LMS, tumor cells may have an epithelial appearance and few mitoses. The diagnostic criteria are met if nuclear atypia is moderate, and either coagulation necrosis or 1.6 or more mitoses per mm^2 are observed. In myxoid LMS, abundant mucinous matrix is seen between tumor cells, and nuclear atypia and mitosis are scarce. For diagnosis, it is essential to note irregular margins and infiltrative growth in the surrounding area.

The prognosis for each stage of the disease is summarized in Table 12.1. The staging system of uLMS differs from that of endometrial cancer based on FIGO2008. In principle, staging is determined based on the FIGO2008 classification upon grasping the degree of tumor progression according to laparotomy findings. According to the tumor registry of the Japan Society of Obstetrics and Gynecology, the distribution of uLMS cases in Japan is 65% for stage I, 10% for stage II, 8% for stage III, and 17% for stage IV [6]. Furthermore, the Japanese Gynecologic Oncology Group (JGOG) 2049s examined the treatment details and prognosis of uLMS in Japan and found out that the disease has a poor prognosis with a 5-y OS rate of 57.1%, 44.6%, 20.4%, and 23.2% for stage I, stage II, stage III, and stage IV, respectively [7].

Table 12.1 The 5-year overall survival rate for each stage of uLMS

FIGO2008	Distribution [6] (%)	5-y OS [7] (%)	Description
Stage I IA IB	65.7	57.1	Tumors localized to the uterus Tumor size ≤5 cm Tumor size >5 cm
Stage II IIA IIB	9.8	44.6	Tumor extending to the pelvic region Invasion to adnexa Invasion to other pelvic tissues
Stage III IIIA IIIB IIIC	7.8	20.4	Tumors extending outside the pelvis One part Two or more parts Metastasis to pelvic lymph nodes and/or para-aortic lymph node
Stage IV IVA IVB	16.7	23.2	Invasion of urinary bladder mucosa and/or rectal mucosa Remote metastasis

12.3 Current Standard Therapy for uLMS

The only effective treatment for uLMS is complete surgical resection. After complete resection, no established adjuvant chemotherapy regimen exists [8], and adjuvant radiation therapy is ineffective [9]. Chemotherapy is the primary treatment option for unresectable advanced or recurrent cases.

For resectable cases, the standard surgical procedures include total abdominal hysterectomy (TAH) and bilateral salpingo-oophorectomy (BSO). However, ovarian resection and lymph node dissection can be omitted in uLMS [10].

- Ovarian metastasis is rare (around 4%), so ovarian preservation may be considered in younger patients who have a significant time before menopause and in cases where the tumor is confined to the uterus.
- Hematogenous metastasis to the lungs and liver can occur early in uLMS.
- Retroperitoneal lymph node metastasis is relatively rare, with rates of 6–11% overall and 0–4% in cases where the tumor is macroscopically confined to the uterus. Therefore, lymph node dissection is generally not considered significant.
- If lymph node metastasis is already present, many cases also exhibit hematogenous metastasis to the lungs or liver or peritoneal dissemination. As a result, lymph node dissection (biopsy) is performed only when lymph node enlargement is confirmed via imaging [11].

In Japan, postoperative adjuvant chemotherapy had been performed in about 60% of post-initial operative cases, including total resection cases. In JGOG2049s, the absence of adjuvant therapy is an independent poor prognostic factor (hazard ratio (HR): 1.62, 95% confidence interval (CI): 1.12–2.34, $p = 0.010$). About 40%

of cases that underwent adjuvant chemotherapy regimens chose gemcitabine-docetaxel combination therapy (GD therapy) [7].

Contrarily, although several meta-analyses for adjuvant chemotherapy for uLMS have been conducted [8], no definitive conclusion for early cases (stages I and II) has been reached regarding the impact of postoperative adjuvant chemotherapy on prognosis. Thus, observation is recommended in early-stage cases, while adjuvant chemotherapy may be considered in advanced cases where complete resection has been performed. While an established adjuvant chemotherapy regimen has yet to be determined, GD therapy or doxorubicin (Dox) monotherapy is considered [11, 12].

12.4 First-Line Chemotherapy for Advanced or Recurrent uLMS

Due to its rarity, systemic chemotherapies at the first line have been developed in clinical trials for advanced or recurrent STS, including LMS. The key drug is the anthracycline agent, Dox, and the efficacy of combination therapy with various alkylating agents has been verified, as shown in Fig. 12.2. No regimen has long been superior to Dox monotherapy regarding OS. However, combination therapy of Dox with trabectedin improved OS compared with Dox monotherapy in a randomized phase 3 trial [13], one of the preferred regimens in NCCN Guidelines version 2.2025 [12].

Table 12.2 summarizes the phase 3 randomized controlled trials (RCTs) of first-line chemotherapy for unresectable recurrent or advanced STS, including uLMS. Although the efficacy of alkylating agent and Dox combination therapy has been compared with Dox monotherapy, alkylating agents, such as

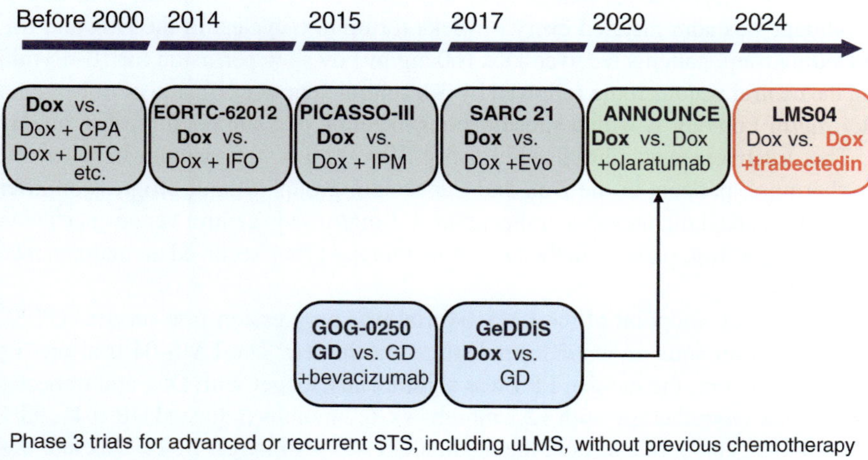

Fig. 12.2 Phase 3 trials at first-line chemotherapy for advanced or recurrent STS, including uLMS. Dox doxorubicin, CPA cyclophosphamide, DTIC dacarbazine, IFO ifosfamide, IPM palifosfamide, EVO evofosfamide, GD gemcitabine-docetaxel

Table 12.2 First-line chemotherapy for advanced or recurrent STS (phase 3 trial, including uLMS)

Regimen	Year	Total N of Pts	N of LMS	N of uLMS	ORR (%)	mPFS (M)	mOS (M)
Dox vs. Dox + CPA [14]	1985	104	38	38	19 vs. 19	5.1 vs. 4.9	11.6 vs. 10.9
Dox + DTIC bolus vs. infusion [15]	1991	240	105	27	17 vs. 17	NA	NA
GD + placebo vs. GD + Bev [16]	2015	107	107	107	31.5 vs. 35.8	6.2 vs. 4.2	26.9 vs. 23.3
Dox vs. GD [17]	2017	257	118	71	19 vs. 20	5.8 vs. 5.9	19.1 vs. 16.8
Dox + placebo vs. Dox + olaratumab [18]	2019	506	234	94	18.3 vs. 14.0	6.8 vs. 5.4	19.7 vs. 20.4
Dox vs. Dox + trabectedin [13, 19]	2022, 2024	150	150	67	15 vs. 36	6.2 vs. 12.2	24 vs. 33

N number, *Pts* patients, *LMS* leiomyosarcoma, *uLMS* uterine leiomyosarcoma, *ORR* objective response rate, *mPFS* median progression-free survival, *mOS* median overall survival, *M* month, *NA* not available, *Dox* doxorubicin, *CPA* cyclophosphamide, *DTIC* dacarbazine, *GD* gemcitabine + docetaxel, *Bev* bevacizumab

cyclophosphamide (CPA) [14], dacarbazine (DTIC) [15], ifosfamide (IFO) [20], palifosfamide (IPM) [21], or evofosfamide (EVO) [22], combined with Dox did not demonstrate improved OS compared with Dox monotherapy, except for trabectedin and Dox combination therapy [19].

In the LMS04 trial [19], the efficacy of Dox and trabectedin combination therapy followed by trabectedin maintenance therapy compared to Dox monotherapy was verified for only patients with LMS. Patients in the Dox monotherapy group received Dox (75 mg/m^2) via the central venous route per slow perfusion for 10–15 min. An injection of subcutaneous lenograstim was given every day from day 3 to day 9. Treatment was administered every 3 weeks for up to 6 cycles. In the Dox and trabectedin group, patients received Dox (60 mg/m^2) by slow perfusion for 10–15 min via the central venous route followed by 3-h central venous perfusion of trabectedin (1.1 mg/m^2) on day 1, with a subcutaneous injection of 6 mg pegfilgrastim on day 2. This treatment was administered every 3 weeks for a maximum of 6 cycles. Furthermore, patients in the Dox and trabectedin group without progression after 6 cycles received maintenance trabectedin (1.1 mg/m^2) via central venous perfusion for 3 h. In contrast, patients in the Dox monotherapy group received no maintenance treatment.

The primary endpoint of the LMS-04 trial was progression-free survival (PFS), validated by an anonymous independent central review. The LMS-04 trial met its primary endpoint; the median PFS was significantly longer with Dox and trabectedin vs. Dox monotherapy with 12.2 months vs. 6.2 months (adjusted HR 0.41, 95% CI 0.29–0.58) [19]. In addition, the median OS was also longer in the Dox and trabectedin group with 33 months than in the Dox monotherapy with 24 months, with the adjusted HR for death of 0.65 (95% CI 0.44–0.95) [13].

When used as a single agent in the second line or later, trabectedin is administered continuously over 24 h at a 1.5 mg/m^2 dose. Still, due to its high hematologic toxicity, the initial dose must be adjusted in some ethnic groups, such as the Japanese [23]. When combined with Dox in the LMS-04 trial, trabectedin was administered at a dose of 1.1 mg over 3 h, with pegfilgrastim support, but even so, 97% of patients experienced grade 3 or 4 hematologic toxicity. In addition, liver dysfunction was observed in 46% of patients. Despite these treatment-related toxicities, 81% of patients in the Dox and trabectedin group received 6 cycles of treatment, and no treatment-related deaths were reported in the group [13].

On the other hand, among molecular targeted drugs, olaratumab, a monoclonal antibody for a subunit alpha of platelet-derived growth factor receptor (PDGFRα), had shown promise. Although randomized phase II trials showed promising results, phase III trials did not demonstrate superior efficacy to Dox monotherapy.

In the ANNOUNCE trial, the additional effect of olaratumab on Dox therapy was verified as an RCT [18]. In a prior phase Ib and a randomized phase 2 trial ($n = 133$), the combination of Dox and olaratumab compared to Dox monotherapy had shown an improvement in prognosis with the median PFS of 6.6 months vs. 4.1 months (stratified HR 0.67; 0.44–1.02, $p = 0.0615$), and with the median OS of 26.5 months vs. 14.7 months (stratified HR 0.46, 0.30–0.71, $p = 0.0003$) [24], resulting in receiving expedited approval by FDA and EMA. However, unexpectedly, efficacy was not verified in phase 3 trials. Based on the results of phase 2 trials, patients in the Dox plus placebo group in the ANNOUNCE trial were treated with Dox 75 mg/m^2 (day 1) and placebo (days 1 and 8), and those in the Dox plus olaratumab group were treated with Dox 75 mg/m^2 (day 1) and olaratumab at 20 mg/kg only in cycle 1 and 15 mg/kg (days 1 and 8) in cycles 2–8 in the combination therapy period (up to 8 cycles), and placebo (days 1 and 8) or olaratumab 15 mg/kg (days 1 and 8) were continued during the maintenance therapy period until progression, respectively. The primary endpoint was OS, conducted on 509 subjects with stricter significance levels and statistical power than the preceding randomized phase 2 trial. In addition, the treatment effect on the LMS population was examined as a planned subgroup analysis. As a result, no statistically significant difference in OS was observed between the Dox plus olaratumab group vs. the Dox plus placebo group in either population (total STS: hazard ratio, 1.05 [95% CI 0.84–1.30], $p = 0.69$, median OS, 20.4 months vs. 19.7 months; LMS: hazard ratio, 0.95 [95% CI 0.69–1.31], $p = 0.76$, median OS 21.6 months vs. 21.9 months) [25].

Furthermore, GD therapy was another treatment option for uLMS because of the high ORR for uLMS in a phase 2 trial [26]. In GOG-0250, a phase 3 RCT was conducted for only patients with uLMS that compared GD therapy and GD therapy + bevacizumab (Bev). The results found no significant difference in terms of OS and PFS, which ruled out the additional effect of Bev [16]. On the other hand, the GeDDiS trial, an RCT on Dox monotherapy, and GD therapy for STS including uLMS did not demonstrate the superiority of GD therapy [17]. Therefore, Dox monotherapy or Dox and trabectedin combination therapy, when toxicity is acceptable, is the standard regimen for uLMS without previous history of chemotherapy.

12.5 Second-Line Chemotherapy for Advanced or Recurrent uLMS

Table 12.3 shows significant RCTs for advanced or recurrent STS, including uLMS with a previous history of chemotherapy. There is no standard regimen for second-line or later chemotherapy for STS, including uLMS that has relapsed after previous chemotherapy. Drugs that have shown efficacy in the primary endpoint in current RCTs include pazopanib, trabectedin, and eribulin.

In a placebo-controlled RCT of patients with STS who had previously been treated with chemotherapy, pazopanib was shown to have more efficacy than the placebo group in terms of PFS. Although the number of uLMS cases was unknown, since LMS is a primary histologic subtype, pazopanib was effective in terms of PFS in the subgroup analysis [27].

In a randomized phase 3 trial comparing trabectedin with DTIC for advanced or recurrent LMS and liposarcoma with a history of Dox monotherapy, trabectedin at a starting dose of 1.5 mg/m^2 was shown to be more effective than DTIC in terms of PFS [30]. In the subgroup analysis for uLMS, trabectedin also showed a significant prolongation in terms of PFS over the DTIC group. However, the incidence of severe myelosuppression is high in Japan. From the domestic phase 1 trial results, the starting dose is 1.2 mg/m^2, a one-step reduced dose in Europe and the United States [23].

Table 12.3 Secondary chemotherapy for advanced or recurrent STS (phase 3 trials)

Line	Regimen	Year	Total N of Pts	N of LMS	N of uLMS	ORR (%)	mPFS (M)	mOS (M)
3rd	**Pazopanib vs. placebo** [27]	**2012**	**369**	**165**	**NA**	**6 vs. 0**	**4.6 vs. 1.6**[a]	**12.5 vs. 10.7**
2nd	Ridaforolimus vs. placebo [28]	2013	711	231	NA	**(CBR: 40.6 vs. 28.6**[a])	**4.4 vs. 3.65**[a]	22.7 vs. 21.3
2nd	Ombrabulin + CDDP vs. placebo + CDDP [29]	2015	355	95	NA	4 vs. 1	**1.54 vs. 1.41**[a]	11.44 vs. 9.33
2nd	**Trabectedin vs. DTIC** [30]	**2016**	**518**	**378**	**212**	**9 vs. 10**	**4.2 vs. 1.5**[a]	**12.4 vs. 12.9**
3rd	**Eribulin vs. DTIC** [31]	**2016**	**452**	**297**	**131**	**4 vs. 5**	**2.6 vs. 2.6**	**13.5 vs. 11.5**[a]

N number, *Pts* patients, *STS* soft tissue sarcoma, *LMS* leiomyosarcoma, *uLMS* uterine leiomyosarcoma, *ORR* objective response rate, *mPFS* median progression-free survival, *mOS* median overall survival, *M* month, *NA* not available, *CBR* clinical benefit rate (proportion of patients achieving CR, PR, or SD for ≧4 months), *CDDP* cisplatin, *DTIC* dacarbazine

[a]$p < 0.05$. The bolded trials showed a new treatment option for STS

Eribulin was the only approved drug to prolong OS in a randomized phase 3 trial, compared with DTIC for advanced or recurrent LMS and liposarcoma with a history of chemotherapy, including Dox monotherapy [31]. In the subgroup analysis for LMS [32], especially for uLMS, the DTIC group exceeded the eribulin group in terms of OS.

Table 12.4 summarizes phase 2 trials for second-line or later conducted for uLMS. Many of these trials were conducted in a single arm, and the primary endpoint was overall response rate (ORR) or PFS rate at 12 weeks or 6 months. In a single-arm trial, caution is required when interpreting the results of time-to-event endpoints such as OS or PFS; however, the GD regimen, letrozole, and trabectedin met the pre-defined efficacy criteria. Furthermore, among RCTs in phase 2 trials beyond second-line chemotherapy targeting only uLMS, the TAXOGEN trial randomly compared gemcitabine monotherapy (GEM therapy) vs. GD therapy. The results did not demonstrate the additional effect of docetaxel in GEM single-agent; ORR was 19% vs. 24%, and median PFS did not exceed 5.5 months vs. 4.7 months [36]. These results highlight the urgent need for the development of new treatments.

Table 12.4 Secondary chemotherapy for advanced or recurrent uLMS (phase 2 trials)

Line	Regimen	Year	Total N of Pts	N of LMS	N of uLMS	ORR (%)	mPFS (M)	mOS (M)
2nd	PTX [33]	2003	48	48	48	8.4	NA	NA
2nd	GEM [34]	2004	42	42	42	20.5	NA	NA
2nd	**GD** [35]	**2008**	**48**	**48**	**48**	**27**[a]	**6.7 M**	**14.7 M**
2nd	GEM vs. GD [36]	2012	90	90	46[b]	19 vs. 24	5.5. M vs. 4.7 M	NA
2nd	lxabepilone [37]	2014	23	23	23	0	1.4 M	NA
2nd	**Letrozole** [38]	**2014**	**27**	**27**	**27**	**0**	**3 M (PFS-12 W 50%**[a]**)**	**NA**
2nd	Nivolumab [39]	2017	12	12	12	0	1.8 M	NA
2nd	Alisertib [40]	2017	23	23	23	0	1.7 M	14.5 M
2nd	**Trabectedin** [41]	**2018**	**168**	**168**	**168**	**23.5%**	**4.1 M (PFS-6 M 35.2%**[a]**)**	**20.6 M**
2nd +	Thalidomide [42]	2007	29	29	29	0%	1.9 M	8.3 M
2nd +	Sunitinib [43]	2009	23	23	23	8.7%	1.54 M	15.1 M

N number, *Pts* patients, *LMS* leiomyosarcoma, *uLMS* uterine leiomyosarcoma, *ORR* objective response rate, *mPFS* median progression-free survival, *mOS* median overall survival, *M* month, *NA* not available, *PTX* paclitaxel, *GEM* gemcitabine, *GD* gemcitabine + docetaxel, *PFS-12 W* progression-free survival rate at 12 weeks after the study entry, *PFS-6 M* progression-free survival rate at 6 months after the study entry

[a]Superior to the alternative rate of primary endpoint in the trial, and the bolded regimens should be developed further in a phase 3 trial

[b]An independent randomized phase 2 trial for uLMS and nuLMS

12.6 Molecular Characteristics of uLMS

From a molecular biological perspective, uLMS is a mesenchymal malignant tumor classified as a complex karyotype. There have been few reports of characteristic fusion genes. Furthermore, unlike many epithelial malignant tumors, there are few gene mutations in tumor tissue, and the molecular biological characteristic is severe chromosomal copy number abnormalities (aneuploidy) [44]. Therefore, it is necessary to explore molecular genetic prognostic factors entirely different from those in epithelial malignant tumors.

In a report of 80 cases of uLMS, the gene mutations in tumor tissue were examined using MSK-IMPACT, a targeted NGS testing. The most frequently observed mutated genes were *TP53*, *RB1*, *ATRX*, *PTEN*, and *BRCA2* with mutation rates of 56%, 51%, 31%, 19%, and 7.5%, respectively [5]. Despite the relatively high rate of *PTEN* loss, which is known to be accompanied by pathogenic mutations in *POLE*, no *POLE* alterations were reported. Only two patients (2.5%) showed an MSI-high in MSIsensor, suggesting that even if the molecular classification recommended for epithelial malignancies of the uterine corpus in the fifth edition of the WHO Classification [3] is applied, subtypes of *POLE* and MMR-deficient are rarely present in uLMS, shown in Fig. 12.3.

Several reports have shown the effectiveness of immune checkpoint inhibitors for MSI-high or tumor-mutation burden (TMB)-high uLMS [45, 46]. On the other hand, in a study comparing treatment-resistant and treatment-sensitive tumors in a uLMS patient who had achieved remission with immune checkpoint inhibitors for more than 2 years, biallelic deletion of *PTEN* and reduced expression of neoantigens highly reactive to patient T cells in vitro were found in treatment-resistant tumors, suggesting that *PTEN* loss is associated with resistance to immune checkpoint inhibitors [47].

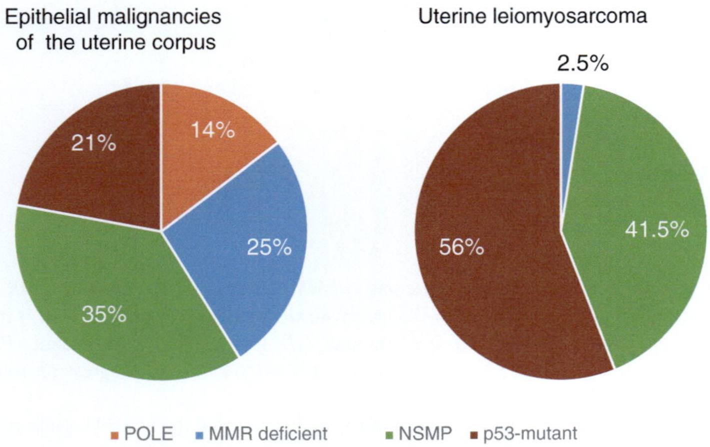

Fig. 12.3 Comparison with the frequency of molecular subtypes of epithelial malignancies of the uterine corpus

Although molecular-targeted drugs are only effective for a few druggable alterations, deletions or pathogenic mutations in the *BRCA2* gene may be suitable for uLMS. A pan-cancer analysis showed that uLMS often has a somatic homozygous deletion of *BRCA2*, a key biomarker for response to poly adenosine diphosphate-ribose polymerase (PARP) inhibitors. In contrast, in other gynecologic cancers like ovarian cancer, *BRCA1/2* inactivation is mainly caused by germline mutations and loss of heterozygosity (LOH) [48].

Studies have found that *BRCA2* mutations are more common in uLMS than non-uterine LMS (~10% vs. 1%) [5, 49]. The frequency of *BRCA1/2* gene mutations in uLMS reported to date ranges from 7.5% to 10%, and HRR-related genes, such as *RAD51B* (9%), *ERCC5* (3%), *RAD51* (2.5%), and *ATR* (1.7%), were present in about 20% of patients [50]. An analysis of 214 STS samples from The Cancer Genome Atlas (TCGA) database assessed homologous recombination deficiency (HRD) scores. It found that an HRD score above 34.5 was linked to worse prognosis, making HRD a promising treatment target for uLMS [51].

For HRD-positive tumors, including those with *BRCA1/2* mutations, PARP inhibitors are being actively researched. Preclinical studies [51, 52] show that PARP inhibitors have antitumor effects not only in ovarian cancer but also in uterine sarcoma cell lines with *BRCA2* mutation. These effects are dose-dependent and enhance cisplatin cytotoxicity. Furthermore, niraparib also showed dose-dependent antitumor activity in soft tissue sarcoma cell lines, including fibrosarcoma and uLMS, both confirmed as HRD-positive. Importantly, these cell lines did not have *BRCA1/2* mutations, suggesting that PARP inhibitors may work in HRD-positive uLMS even without *BRCA1/2* mutations [51, 53].

Table 12.5 summarizes clinical studies on PARP inhibitors for uLMS patients. Research has shown positive outcomes in patients with *BRCA2* alterations.

- In one study, four patients (three with deletions, one with a truncating mutation) received treatment. Among them, one had a partial response (PR), while three showed stable disease (SD) for over 12 months [49].
- Another study examined six patients (four with deletions, two with truncating mutations + LOH). This study reported one complete response (CR) and one PR, with treatment durations ranging from 6 to 28 months [5].

Table 12.5 Therapeutic effects of PARP inhibitors on uLMS with *BRCA2* deletions or truncating mutations

Author	Year	N of Pts	Line of therapy	Response
Seligson [49]	2019	4	Third: 2 (50%) Fourth: 1 (25%) Fifth: 1 (25%)	PR: 1 (25%) SD: 3 (75%)
Hensley [5]	2020	6	NA	CR: 1 (17%) PR: 1 (17%) All had at least some radiographic regression

CR complete response, *PR* partial response, *SD* stable disease, *NA* not available

Both preclinical and clinical data suggest that PARP inhibitors may help prolong OS in HRD-positive uLMS patients, even if CR is not achieved, as long as their condition remains stable.

12.7 Future Directions

The prognosis of uLMS is poor, and the effect of drug therapy is limited, especially for advanced and recurrent cases with unresectable lesions. On the other hand, with the development of genomic analysis technology, it has been found that uLMS has specific genomic alterations. Furthermore, it has been reported that drugs that are expected to be effective can be predicted by chromosomal instability signatures based on whole genome copy number analysis in pan-cancer [54]. Because uLMS is characterized by tumor aneuploidy, it is thought that treatment methods will need to be optimized by subtype classification based on the degree of chromosomal instability, not only gene mutations. Thus, in the future, developing novel therapies based on genome alterations as precision medicine is necessary. In the confirmatory clinical trials conducted so far, due to their rarity, it has been difficult to conduct randomized trials for uLMS alone, and randomized trials have been undertaken with other sarcomas. Securing sufficient subjects in clinical trials to develop therapies based on genomic abnormalities is expected to be even more difficult. Therefore, it is thought that new strategies, such as a method to verify the effectiveness of treatment using large-scale registry studies as external controls and a system for accumulating genomic alterations and clinical information, will be required.

References

1. Toro JR, et al. Incidence patterns of soft tissue sarcomas, regardless of primary site, in the surveillance, epidemiology and end results program, 1978-2001: an analysis of 26,758 cases. Int J Cancer. 2006;119:2922–30. https://doi.org/10.1002/ijc.22239.
2. Siegel RL, Kratzer TB, Giaquinto AN, Sung H, Jemal A. Cancer statistics, 2025. CA Cancer J Clin. 2025;75:10–45. https://doi.org/10.3322/caac.21871.
3. WHO Classification of Tumors Editorial Board. Female genital tumours, WHO Classification of Tumors, vol. 4. 5th ed; World Health Organization, 2020.
4. Chiang S, et al. PGR gene fusions identify a molecular subset of uterine epithelioid Leiomyosarcoma with Rhabdoid features. Am J Surg Pathol. 2019;43:810–8. https://doi.org/10.1097/PAS.0000000000001239.
5. Hensley ML, et al. Genomic landscape of uterine sarcomas defined through prospective clinical sequencing. Clin Cancer Res. 2020; https://doi.org/10.1158/1078-0432.CCR-19-3959.
6. Gynecologic Oncology Committee JSOG. Patient annual report for 2016–2021. https://www.jsog.or.jp/medical/624/#databook. Accessed 16 Feb 2025.
7. Takehara K, et al. Clinical status and prognostic factors in Japanese patients with uterine leiomyosarcoma. Gynecol Oncol. 2020;157:115–20. https://doi.org/10.1016/j.ygyno.2020.01.022.
8. Chae SH, et al. Effect of adjuvant therapy on the risk of recurrence in early-stage leiomyosarcoma: a meta-analysis. Gynecol Oncol. 2019;154:638–50. https://doi.org/10.1016/j.ygyno.2019.07.001.

9. Reed NS, et al. Phase III randomised study to evaluate the role of adjuvant pelvic radiotherapy in the treatment of uterine sarcomas stages I and II: an European Organisation for Research and Treatment of Cancer gynaecological cancer group study (protocol 55874). Eur J Cancer (Oxford, England : 1990). 2008;44:808–18. https://doi.org/10.1016/j.ejca.2008.01.019.
10. Seagle BL, et al. Prognosis and treatment of uterine leiomyosarcoma: a National Cancer Database study. Gynecol Oncol. 2017;145:61–70. https://doi.org/10.1016/j.ygyno.2017.02.012.
11. JSOG O. Japan Society of Gynecologic oncology 2023 guidelines for treatment of uterine body neoplasms. 5th ed. Kanehara & Co., Ltd; 2023.
12. National_Comprehensive_Cancer_Network. NCCN clinical practice guidelines in oncology (NCCN guidelines®) uterine neoplasms version 2.2025.
13. Pautier P, et al. Doxorubicin-trabectedin with trabectedin maintenance in leiomyosarcoma. N Engl J Med. 2024;391:789–99. https://doi.org/10.1056/NEJMoa2403394.
14. Muss HB, et al. Treatment of recurrent or advanced uterine sarcoma. A randomized trial of doxorubicin versus doxorubicin and cyclophosphamide (a phase III trial of the Gynecologic Oncology Group). Cancer. 1985;55:1648–53. https://doi.org/10.1002/1097-0142(19850415)55:8<1648::aid-cncr2820550806>3.0.co;2-7.
15. Zalupski M, et al. Phase III comparison of doxorubicin and dacarbazine given by bolus versus infusion in patients with soft-tissue sarcomas: a Southwest Oncology Group Study. J Natl Cancer Inst. 1991;83:926–32. https://doi.org/10.1093/jnci/83.13.926.
16. Hensley ML, et al. Randomized phase III trial of gemcitabine plus docetaxel plus bevacizumab or placebo as first-line treatment for metastatic uterine leiomyosarcoma: an NRG Oncology/Gynecologic Oncology Group Study. J Clin Oncol. 2015;33:1180–5. https://doi.org/10.1200/jco.2014.58.3781.
17. Seddon, B. *et al.* Gemcitabine and docetaxel versus doxorubicin as first-line treatment in previously untreated advanced unresectable or metastatic soft-tissue sarcomas (GeDDiS): a randomised controlled phase 3 trial. Lancet Oncol 18, 1397–1410 (2017). doi: https://doi.org/10.1016/s1470-2045(17)30622-8.
18. Tap WD, et al. Effect of Doxorubicin Plus Olaratumab vs Doxorubicin Plus Placebo on Survival in Patients With Advanced Soft Tissue Sarcomas: The ANNOUNCE Randomized Clinical Trial. JAMA. 2020;323:1266–76. https://doi.org/10.1001/jama.2020.1707.
19. Pautier P, et al. Doxorubicin alone versus doxorubicin with trabectedin followed by trabectedin alone as first-line therapy for metastatic or unresectable leiomyosarcoma (LMS-04): a randomised, multicentre, open-label phase 3 trial. Lancet Oncol. 2022;23:1044–54. https://doi.org/10.1016/S1470-2045(22)00380-1.
20. Judson I, et al. Doxorubicin alone versus intensified doxorubicin plus ifosfamide for first-line treatment of advanced or metastatic soft-tissue sarcoma: a randomised controlled phase 3 trial. Lancet Oncol. 2014;15:415–23. https://doi.org/10.1016/S1470-2045(14)70063-4.
21. Ryan CW, et al. PICASSO III: a phase III, placebo-controlled study of doxorubicin with or without palifosfamide in patients with metastatic soft tissue sarcoma. J Clin Oncol. 2016;34:3898–905. https://doi.org/10.1200/JCO.2016.67.6684.
22. Tap WD, et al. Doxorubicin plus evofosfamide versus doxorubicin alone in locally advanced, unresectable or metastatic soft-tissue sarcoma (TH CR-406/SARC021): an international, multicentre, open-label, randomised phase 3 trial. Lancet Oncol. 2017;18:1089–103. https://doi.org/10.1016/S1470-2045(17)30381-9.
23. Ueda T, et al. Phase I and pharmacokinetic study of trabectedin, a DNA minor groove binder, administered as a 24-h continuous infusion in Japanese patients with soft tissue sarcoma. Investig New Drugs. 32, 2014:691–9. https://doi.org/10.1007/s10637-014-0094-5.
24. Tap WD, et al. Olaratumab and doxorubicin versus doxorubicin alone for treatment of soft-tissue sarcoma: an open-label phase 1b and randomised phase 2 trial. Lancet. 2016;388:488–97. https://doi.org/10.1016/s0140-6736(16)30587-6.
25. Tap WD, et al. Effect of doxorubicin plus olaratumab vs doxorubicin plus placebo on survival in patients with advanced soft tissue sarcomas: the ANNOUNCE randomized clinical trial. JAMA. 2020;323:1266–76. https://doi.org/10.1001/jama.2020.1707.

26. Hensley ML, Blessing JA, Mannel R, Rose PG. Fixed-dose rate gemcitabine plus docetaxel as first-line therapy for metastatic uterine leiomyosarcoma: a Gynecologic Oncology Group phase II trial. Gynecol Oncol. 2008;109:329–34. https://doi.org/10.1016/j.ygyno.2008.03.010.
27. van der Graaf WT, et al. Pazopanib for metastatic soft-tissue sarcoma (PALETTE): a randomised, double-blind, placebo-controlled phase 3 trial. Lancet (London, England). 2012;379:1879–86. https://doi.org/10.1016/s0140-6736(12)60651-5.
28. Demetri GD, et al. Results of an international randomized phase III trial of the mammalian target of rapamycin inhibitor ridaforolimus versus placebo to control metastatic sarcomas in patients after benefit from prior chemotherapy. J Clin Oncol. 2013;31:2485–92. https://doi.org/10.1200/jco.2012.45.5766.
29. Blay JY, et al. Ombrabulin plus cisplatin versus placebo plus cisplatin in patients with advanced soft-tissue sarcomas after failure of anthracycline and ifosfamide chemotherapy: a randomised, double-blind, placebo-controlled, phase 3 trial. Lancet Oncol. 2015;16:531–40. https://doi.org/10.1016/s1470-2045(15)70102-6.
30. Demetri GD, et al. Efficacy and safety of trabectedin or dacarbazine for metastatic liposarcoma or leiomyosarcoma after failure of conventional chemotherapy: results of a phase III randomized Multicenter clinical trial. J Clin Oncol. 2016;34:786–93. https://doi.org/10.1200/jco.2015.62.4734.
31. Schoffski P, et al. Eribulin versus dacarbazine in previously treated patients with advanced liposarcoma or leiomyosarcoma: a randomised, open-label, multicentre, phase 3 trial. Lancet (London, England). 2016;387:1629–37. https://doi.org/10.1016/s0140-6736(15)01283-0.
32. Blay JY, et al. Eribulin versus dacarbazine in patients with leiomyosarcoma: subgroup analysis from a phase 3, open-label, randomised study. Br J Cancer. 2019;120:1026–32. https://doi.org/10.1038/s41416-019-0462-1.
33. Gallup DG, Blessing JA, Andersen W, Morgan MA, Gynecologic Oncology Group Study. Evaluation of paclitaxel in previously treated leiomyosarcoma of the uterus: a gynecologic oncology group study. Gynecol Oncol. 2003;89:48–51. https://doi.org/10.1016/s0090-8258(02)00136-1.
34. Look KY, et al. Phase II trial of gemcitabine as second-line chemotherapy of uterine leiomyosarcoma: a Gynecologic Oncology Group (GOG) Study. Gynecol Oncol. 2004;92:644–7. https://doi.org/10.1016/j.ygyno.2003.11.023.
35. Hensley ML, et al. Fixed-dose rate gemcitabine plus docetaxel as second-line therapy for metastatic uterine leiomyosarcoma: a Gynecologic Oncology Group phase II study. Gynecol Oncol. 2008;109:323–8. https://doi.org/10.1016/j.ygyno.2008.02.024.
36. Pautier P, et al. Randomized multicenter and stratified phase II study of gemcitabine alone versus gemcitabine and docetaxel in patients with metastatic or relapsed leiomyosarcomas: a Federation Nationale des Centres de Lutte Contre le Cancer (FNCLCC) French Sarcoma Group Study (TAXOGEM study). Oncologist. 2012;17:1213–20. https://doi.org/10.1634/theoncologist.2011-0467.
37. Duska LR, et al. A phase II evaluation of ixabepilone (IND #59699, NSC #710428) in the treatment of recurrent or persistent leiomyosarcoma of the uterus: an NRG Oncology/Gynecologic Oncology Group Study. Gynecol Oncol. 2014;135:44–8. https://doi.org/10.1016/j.ygyno.2014.07.101.
38. George S, et al. Phase 2 trial of aromatase inhibition with letrozole in patients with uterine leiomyosarcomas expressing estrogen and/or progesterone receptors. Cancer. 2014;120:738–43. https://doi.org/10.1002/cncr.28476.
39. Ben-Ami E, et al. Immunotherapy with single agent nivolumab for advanced leiomyosarcoma of the uterus: results of a phase 2 study. Cancer. 2017;123:3285–90. https://doi.org/10.1002/cncr.30738.
40. Hyman DM, et al. A phase 2 study of alisertib (MLN8237) in recurrent or persistent uterine leiomyosarcoma: an NRG Oncology/Gynecologic Oncology Group Study 0231D. Gynecol Oncol. 2017;144:96–100. https://doi.org/10.1016/j.ygyno.2016.10.036.

41. Gadducci A, et al. A phase II randomised (calibrated design) study on the activity of the single-agent trabectedin in metastatic or locally relapsed uterine leiomyosarcoma. Br J Cancer. 2018;119:565–71. https://doi.org/10.1038/s41416-018-0190-y.
42. McMeekin DS, et al. A phase II trial of thalidomide in patients with refractory leiomyosarcoma of the uterus and correlation with biomarkers of angiogenesis: a Gynecologic Oncology Group Study. Gynecol Oncol. 2007;106:596–603. https://doi.org/10.1016/j.ygyno.2007.05.013.
43. Hensley ML, et al. Sunitinib malate in the treatment of recurrent or persistent uterine leiomyosarcoma: a Gynecologic Oncology Group phase II study. Gynecol Oncol. 2009;115:460–5. https://doi.org/10.1016/j.ygyno.2009.09.011.
44. Cancer Genome Atlas Research Network. Electronic address, e. d. s. c. & Cancer Genome Atlas Research N. Comprehensive and integrated genomic characterization of adult soft tissue sarcomas. Cell. 2017;171:950–965.e928. https://doi.org/10.1016/j.cell.2017.10.014.
45. Wang YJ, et al. Use of pembrolizumab in MSI-high uterine leiomyosarcoma; a case report and review of the literature. Gynecol Oncol Rep. 2021;35:100701. https://doi.org/10.1016/j.gore.2021.100701.
46. Lin R, Muaadz A, Gwo Fuang H. A promising response of refractory uterine leiomyosarcoma to pembrolizumab and temozolomide combination therapy. Gynecol Oncol Rep. 2024;51:101326. https://doi.org/10.1016/j.gore.2024.101326.
47. George S, et al. Loss of PTEN is associated with resistance to anti-PD-1 checkpoint blockade therapy in metastatic uterine Leiomyosarcoma. Immunity. 2017;46:197–204. https://doi.org/10.1016/j.immuni.2017.02.001.
48. Jonsson P, et al. Tumour lineage shapes BRCA-mediated phenotypes. Nature. 2019;571:576–9. https://doi.org/10.1038/s41586-019-1382-1.
49. Seligson ND, et al. BRCA1/2 functional loss defines a targetable subset in Leiomyosarcoma. Oncologist. 2019;24:973–9. https://doi.org/10.1634/theoncologist.2018-0448.
50. Rosenbaum E, et al. Clinical outcome of leiomyosarcomas with somatic alteration in homologous recombination pathway genes. JCO Precis Oncol. 2020;4 https://doi.org/10.1200/PO.20.00122.
51. Li H, et al. Molecular signatures of BRCAness analysis identifies PARP inhibitor Niraparib as a novel targeted therapeutic strategy for soft tissue sarcomas. Theranostics. 2020;10:9477–94. https://doi.org/10.7150/thno.45763.
52. Chudasama P, et al. Integrative genomic and transcriptomic analysis of leiomyosarcoma. Nat Commun. 2018;9:144. https://doi.org/10.1038/s41467-017-02602-0.
53. Dall G, et al. Targeting homologous recombination deficiency in uterine leiomyosarcoma. J Exp Clin Cancer Res. 2023;42:112. https://doi.org/10.1186/s13046-023-02687-0.
54. Drews RM, et al. A pan-cancer compendium of chromosomal instability. Nature. 2022;606:976–83. https://doi.org/10.1038/s41586-022-04789-9.

MIX
Papier aus verantwortungsvollen Quellen
Paper from responsible sources
FSC® C105338

If you have any concerns about our products,
you can contact us on
ProductSafety@springernature.com

In case Publisher is established outside the EU,
the EU authorized representative is:
Springer Nature Customer Service Center GmbH
Europaplatz 3, 69115 Heidelberg, Germany

Printed by Libri Plureos GmbH
in Hamburg, Germany